Foundation Geograph

Industry

Nigel Price
Colin Budge

Edward Arnold

First published 1983
by Edward Arnold (Publishers) Ltd.,
41 Bedford Square, London WC1B 3DQ

British Library Cataloguing in Publication Data
Price, Nigel
    Industry — (Foundation geography)
    1. Industry
    I. Title   II. Series
    338   HC59

    ISBN 0-7131-0664-6

Acknowledgements

The Publishers would like to thank the following for
their permission to reproduce copyright illustrations:

Shell Photographic Service, Fig. 21
Popperfoto, Figs. 2c, 31 and r, 6
United Nations, Fig. 2r (F.A.O.)
National Coal Board, Fig. 3c
Financial Times, Fig. 7
Belgian National Tourist Office, Fig. 15
Hoover Ltd., Fig. 20
The Swedish Institute, Foto Börje Rönnberg,
    Figs. 33, 34, 35, 36
English China Clays Group, Figs. 38, 43 and 44
Swedish National Board of Forestry, Figs. 55 and 56
B.S.C., Figs. 60, 61, 65 and 69
Ford Ltd., Figs. 70, 71 and 72
Wiggins Teape Group Ltd., Fig. 79
Anthony McAroy, Figs. 82 and 85
Syndication Internation, Figs. 86 and 88
Birmingham Post, Fig. 87
Birds Eye, Figs. 93 and 94
Ardal Verk, Figs. 103 and 106
Spanish National Tourist Office, Fig. 117
Swiss National Tourist Office, Figs. 118 and 119
Barnabys Picture Library, Fig. 112

Figs. 14, 46, 47, 49, 105, 110, 111, 115 and 116 have
been supplied by the author.

Set in IBM 11pt Press Roman by 🅰 Tek-Art,
Croydon, Surrey.
Printed in Great Britain by Thomson Litho Ltd.,
East Kilbride.

Contents

1

Introduction

1A What is industry?

Fig. 1 shows some headline cuttings taken from just one newspaper, over only a few days. They show how important industry is in a country such as Britain. The cuttings also show how varied industry is. *Industry* includes all that people do in their working lives. This means that it is more than just making things. It includes finding materials, providing money, organising, buying, selling and supplying our needs. Industry is all types of organised work.

1 Look at Fig. 1 and make a list of all the different industries you can pick out from the cuttings.
2 List all the words that you can find in Fig. 1 that are to do with two important parts of industry, (a) *workers* and (b) *money*.
Say what you think each listed word means.

The word industry, then, covers a very wide range of activities and we need to look more closely at the different types of industry. Fig. 2 shows six types of industry. Below the sketch are six small photographs of workers in these industries.

3 Make a copy or tracing of Fig. 2 and complete it by writing in the name of each industry chosen from the photographs.

Fig. 1

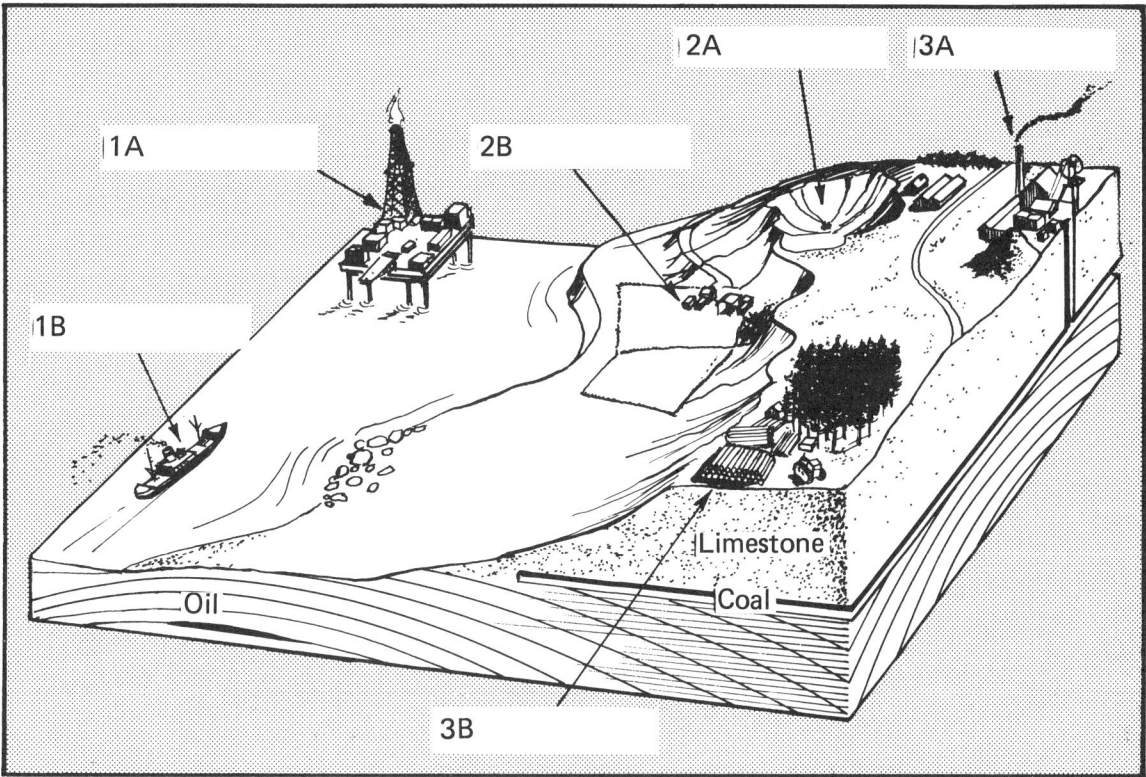

1A

2A

3A

2B

1B

Limestone

Oil

Coal

3B

Fig. 2

Oil drilling

Fishing

Farming

Quarrying

Mining

Forestry

Fig. 3 Industrial workers

1B Primary and secondary industry

All the industries shown in Fig. 2 have something in common; they are all to do with things that nature provides for us, for example, fish from the sea, coal from the ground. When man uses things that occur *naturally* in the sea or on the land, these things are called *natural resources.* Industries such as fishing, quarrying and farming which make use of natural resources are called **primary industries**. Primary industries are important because they provide many **raw materials** needed by other industries — oil for making plastics, wood for making paper and so on.

Primary industries fall into two groups. Look again at Fig. 2. Once the raw materials of group A have been used up the industries have to stop. These resources are *non-renewable.* With group B the raw materials can be used and replaced. These resources are *renewable.*

> 1 Copy and complete Fig. 4 using the industries shown in Fig. 2.

Group A Non-renewable	Group B Renewable
1	
2	
3	

Fig. 4 Primary industries

Most of the raw materials produced by primary industries are then used for making or **manufacturing** other products, for example the timber produced in forestry is manufactured into furniture or paper. The industry that turns raw materials into other useful **end-products** is called manufacturing or **secondary industry**. Secondary industries usually take place in *factories.*

> 2 In your own words explain what is meant by the following: natural resource, raw material, end product, factory.
> 3 Write a paragraph to explain the difference between primary industries and secondary industries. Also explain how the two are connected.

Fig. 5 shows how a typical small factory, making plastic kitchenware, might be organised.

> 4 Make a copy of the diagram.
> 5 What raw materials does the factory use, and what does one load cost?
> 6 Explain what happens to the raw materials during the manufacturing process.
> 7 How much does it cost to manufacture the end-product? How is this cost made up?
> 8 What is the total cost of raw materials and manufacturing?
> 9 How much is the end-product sold for?
> 10 The difference between the factory's costs and the selling price of the end-product is called the factory's *profit*. How much profit is made on this batch of goods?
> *11 Explain what is meant by the words **inputs** and **outputs** that are used on the diagram.
> *12 Why must the inputs always be of lower value than the outputs? What would happen to the factory if they were not?

Figs. 6 and 7 show how secondary industry can be divided. The man on the left, helping to make a ship, is working in **heavy industry**. The woman on the right, putting together a television set, is working in **light industry**. As you can see, heavy industry is to do with making large-scale end-products and handling bulky and heavy raw materials. Light industry uses small amounts of easily handled raw materials to make its end-products.

> 13 Sort out the industries listed in Fig. 8 into two groups: *Heavy* and *Light Industry*.

> Shipbuilding, Television assembly, Girder making, Locomotive building, Fruit canning, Camera making, Clothes manufacture, Brick making

Fig. 8

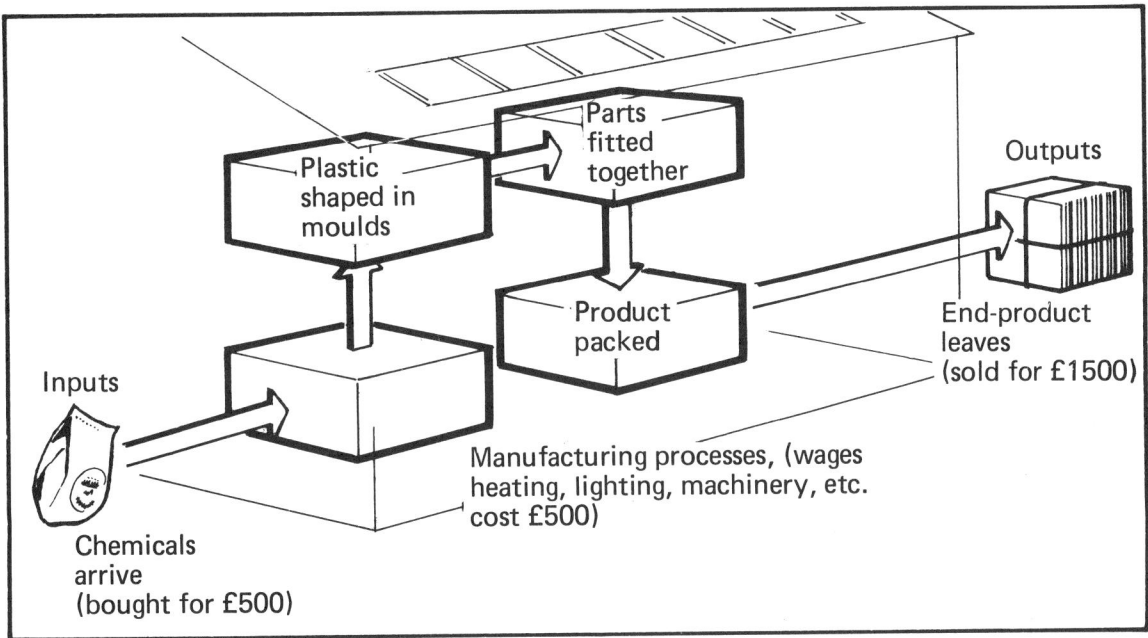

Fig. 5

Fig. 6

Fig. 7

1C Service industries

Not all industries make their profit by selling end-products. In our lives we need other things besides tinned food, clothes and cars.

We need electricity to light our homes and streets, buses and trains to move us round the country, doctors when we are ill, the police to protect us and so on. Here no end-product is made. This third group of industries gives us a *service* — something we need done or provided. *Social services* are those paid for by our *taxes* and *rates*. This group includes the police and the National Health Service. All these services are part of *tertiary industry*.

1 Make a list of the High Street services shown in Fig. 9.
2 Make another list of any services not shown in the sketch but which you would expect to find in a normal High Street.
3 Sort out the industries shown in Fig. 10 into three lists: *Primary, Secondary* and *Tertiary*. There are five in each list.

Fig. 11 shows that when you buy a loaf of bread all three groups of industry have been involved.

4 Make a copy of the diagram, and in your own words explain the part played by primary, secondary and tertiary industries.

Cement making, Supermarket, Window cleaning, Shoe manufacture, Steel making, Growing rubber trees, Diamond mining, Furniture making, Milk production, Garage, Television repairs, Army, Limestone quarrying, Making car tyres, Forestry.

Fig. 10

*5 Some other tertiary industries provide services to the flour mills and bakeries. Suggest some services needed by the secondary industries in the example shown.

Fig. 12 shows the divisions of the workforce employed in Britain's industry. This circular diagram is known as a *pie graph*. It is useful for showing figures as percentages (%); 1% is 1/100 part of the whole quantity shown by the circle, in this case the workforce. We measure the angle at the centre of the circle for each division in degrees (°) with a protractor. There are 360° in a circle, and the circle represents 100%, so:

100% is represented by 360°

1% is represented by $\frac{360}{100} = 3.6°$

Fig. 9

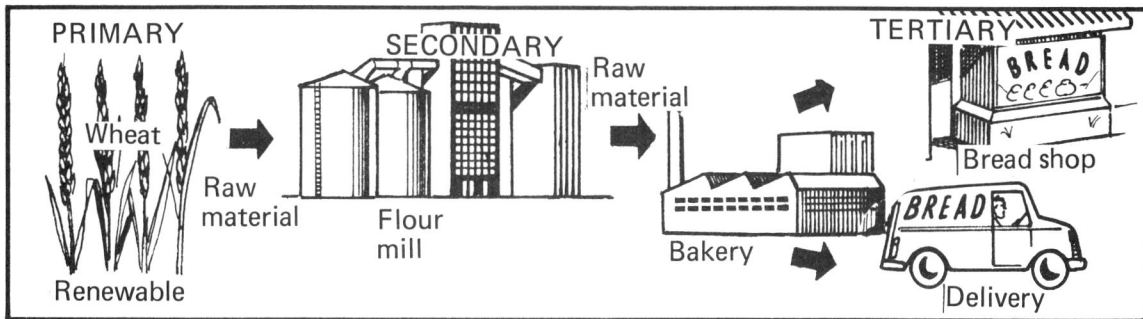

Fig. 11 Bread-making — from field to table

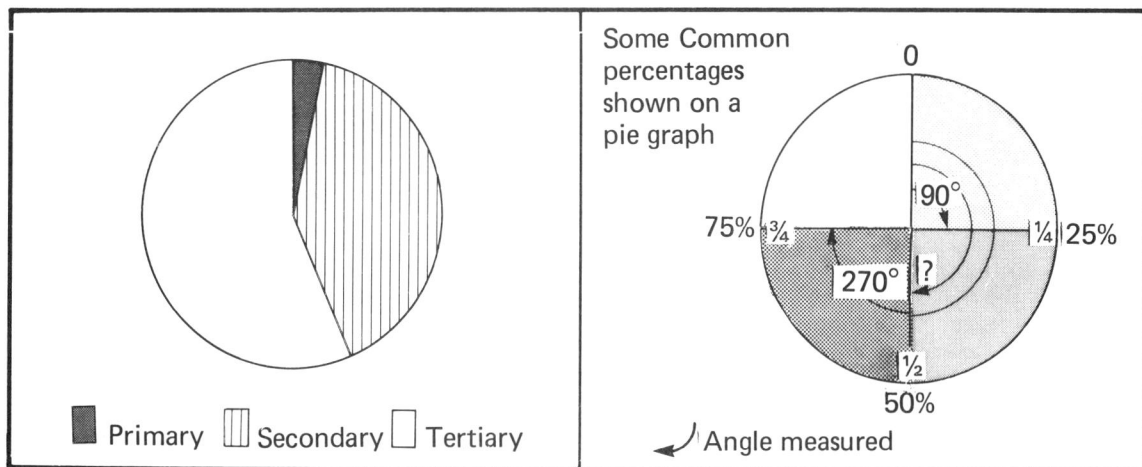

Primary Secondary Tertiary

Some Common percentages shown on a pie graph

Angle measured

Fig. 12 Britain's industrial workforce

6 Fig. 12 also shows how some simple percentages look as parts of a circle. A quarter, or 25%, is shown as 90°; three quarters, or 75%, is shown by 270° of the whole circle. How many degrees are there in half the circle, or 50%?

7 Use a protractor to measure the angles of the three divisions in the workforce circle using the information in the above equation. Work out the percentages of people employed in each division.

8 Use the information from question 7 to complete the following sentences.

_____ % of Britain's workforce is in primary industry. _____ times more workers are employed in secondary industry than in primary industry. Tertiary industry employs most people, making up _____ % of the total workforce. These figures for Britain are typical of the industrial countries of Europe.

Sweden, in northern Europe, is one of the world's most advanced industrial countries.

		% Workers	
		1960	1980
A	Primary industry	25	11
B	Secondary industry	29	28
C	Tertiary industry	46	61

Fig. 13 Sweden's workforce

Fig. 13 shows how the workforce in the three areas of Sweden's industry changed between 1960 and 1980. This sort of change is typical of many European countries.

9 Show the figures as two separate pie graphs, one for the three parts in 1960 and the other for the three parts in 1980.

10 Describe the changes that have taken place.

*11 Why do you think that tertiary industries now employ so many more people than primary or secondary industry? Do you think this is an advantage or disadvantage?

9

1D The start of modern industry

Fig. 14 shows a clockmaking and carpentry works in use today in the Black Forest region of Germany. It shows how production was organised about 200 years ago before the *Industrial Revolution*.

1 What source of power is being used to turn the simple machinery?
2 What machines might the carpenter use in these works?
3 Which other natural sources of power were used by other early industries?
*4 Draw your own labelled diagram to show how this power is taken into the carpentry works.
*5 Why do you think these small industries were widely scattered over a large area, rather than concentrated in just a few places?

This sort of industry is called *cottage industry* because it is on such a small scale and is often carried out in the workers' homes. Industry was on this small scale because the power was unreliable and could only drive a few, simple machines.

Another example of a craft cottage industry is shown in Fig. 15. This is lacemaking in the town of Bruges, in Belgium. The town has been famous for lacemaking for hundreds of years. Most lace is now sold to tourists.

6 Describe how this skilled worker is making the lace.
*7 Say why you think hand-made lace is so expensive.

Fig. 14

Fig. 15

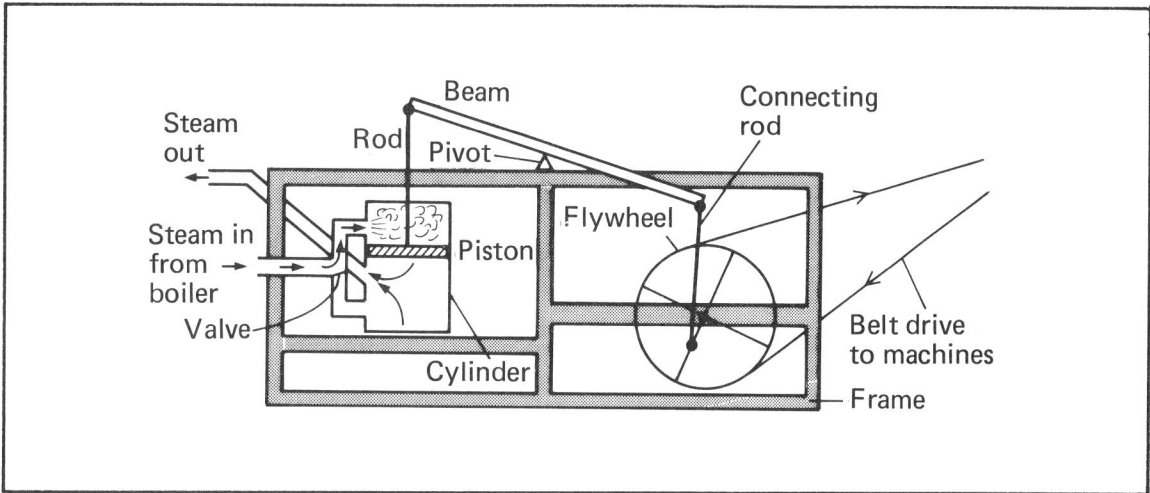

Fig. 16

The most important change in industry came about with the development of the *steam-engine* about 200 years ago. It was the steam-engine that caused the change from cottage industries like the ones in the photographs to the large factories that are common today. This was because the steam-engine was powerful enough to drive many machines at once. The steam-engines burned coal to heat water into steam. The steam, under pressure, pushed a piston to drive wheels and belts. With these new industrial machines workers could now be brought together in large buildings called *factories.* Large factories meant a big need for more industrial workers and this in turn meant that new industrial towns quickly grew up. This change in how industry was carried out is known as the *Industrial Revolution.*

8 Explain why steam-engines gave rise to a change from small scale cottage industry to large scale factory industry.

*9 What effect did the Industrial Revolution have on where industries were located?

10 *Project idea*
    Use the books in the library to find out what you can about *James Watt* and the development of the steam-engine.

*11 Fig. 16 shows an early steam-engine. Make a copy of the diagram and write a paragraph to describe how it worked.

Fig. 17 shows an area of country as it might have been about 150 years ago during the Industrial Revolution. There are five possible sites for a new factory, marked A–E.

12 Which site would you choose for the factory? Give at least four reasons for your choice.

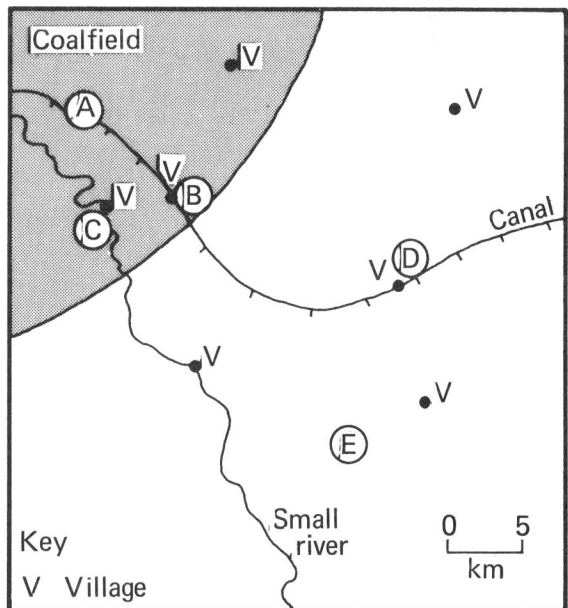

Fig. 17

1E Modern factories

Coal is *bulky* and expensive to transport over long distances. It was the major power source of the Industrial Revolution. Since then factories and towns have grown on Europe's important coalfields. Today these coalfields are places where millions of people live and work in closely clustered towns and cities.

1 Make a copy of Fig. 18 which shows western Europe's coalfields. Complete it by naming the towns and countries shown by their first letters.

Today steam power has been replaced by electric power, and there is a much wider choice of where to build a factory because of this. Despite the use of flexible electric power, coalfields still remain important centres of modern industry. This is because, with their early industrial start, much money has been invested in land, buildings, machinery and communications over the years. Local people have also become skilled workers.

These are powerful reasons for industry to continue in the area where it started. This tendency for industry to stay in the same area is called *industrial inertia*.

2 Explain, in a few sentences, why electricity is a much more flexible source of factory power than coal.
3 Explain what you understand by the term 'industrial inertia'.

A factory is a very complex piece of organisation. Many of the needs involved in the running of a factory are set out in Fig. 19.

4 Make a copy of Fig. 19.
5 Make a list of factory *inputs* and *outputs*. Refer to the section on primary and secondary industry and the *Glossary*, then explain what you understand by the terms inputs and outputs.

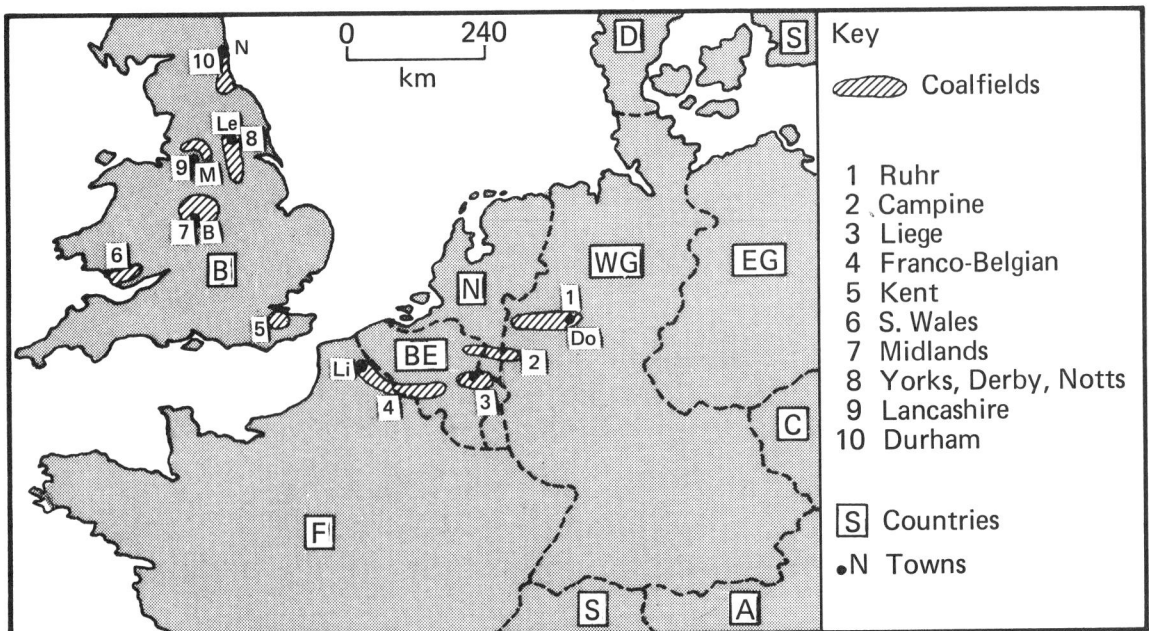

Fig. 18

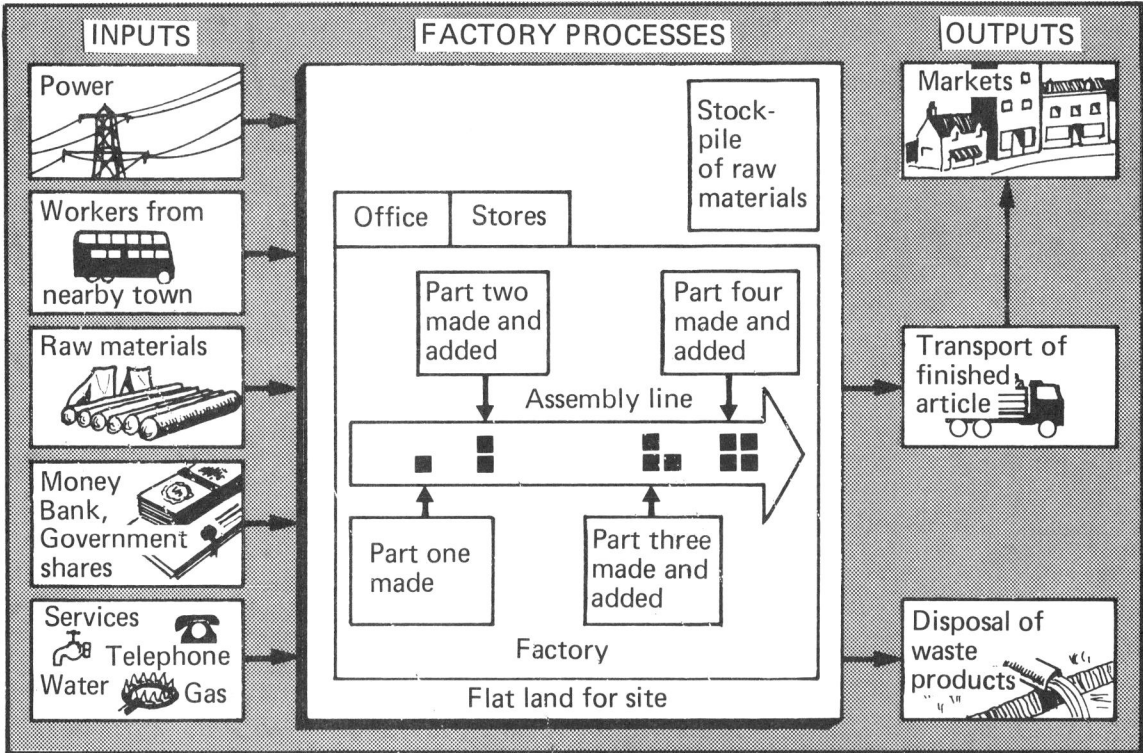

Fig. 19

The worker in the Black Forest workshop we saw makes each article from start to finish. Each article is slightly different. Modern factories do not use this method of production for they turn out many articles, exactly the same, by the process of ***mass production.*** Each group of workers makes a small, different part of the finished article. The many small parts are then joined together on an ***assembly-line*** to complete the end-product. This method increases production because it is faster and each worker only has to have a few skills.

6 In your own words describe what you understand by *mass production* and *assembly line*. List any examples of assembly-line mass production that you can think of.

7 Fig. 20 shows a worker on an assembly line. Describe the work being done.

8 Why does mass production mean cheaper goods?

*9 Write a few sentences about some of the disadvantages of assembly-line mass production.

Fig. 20

2

Obtaining the resources

2A Coalmining

'When I arrive at work I enter the locker room (1) near the canteen (2). Next to the locker room are the baths (3) and the lamp room (4). In the locker room we change into our working clothes before collecting our lamps and going to the pithead at the top of the shaft (5). Alongside lies a second set of headgear above the shaft which brings the coal to the surface (6). At the bottom of the shaft (7) we enter a train with open cars to travel to the coal-face (8). At the coal-face we man a coal-cutting machine which cuts the coal and places it on a conveyor belt (9) which returns it to the bottom of the second shaft from there it is taken to the surface.'

Coalminer

Fig. 21 shows a simplified section of the workings of a coal-mine.

1 Make a copy of the diagram. Use the information to add a key to your drawing explaining the features at numbers (1) to (9).

2 Coal is found in seams in the rocks underground. Label the *coal seam* on your diagram.

3 On your diagram use arrows of different colours to show the routes taken by the coal-miners and the coal. Add these to your key.

*4 At point (A) lies an abandoned coal-mine. This was called a *drift-mine.* How is this mine different from the modern shaft mine and why would this type of mine have been used before a deep shaft?

*5 Look at points (B) and (C). At point (B) there is a *geological fault* which has moved the coal seam. At (C) the seam thins out. Explain why mining this kind of seam is difficult.

*6 *Project ideas*
Using suitable reference books write a short account of a) the formation of coal b) safety in a coal-mine.

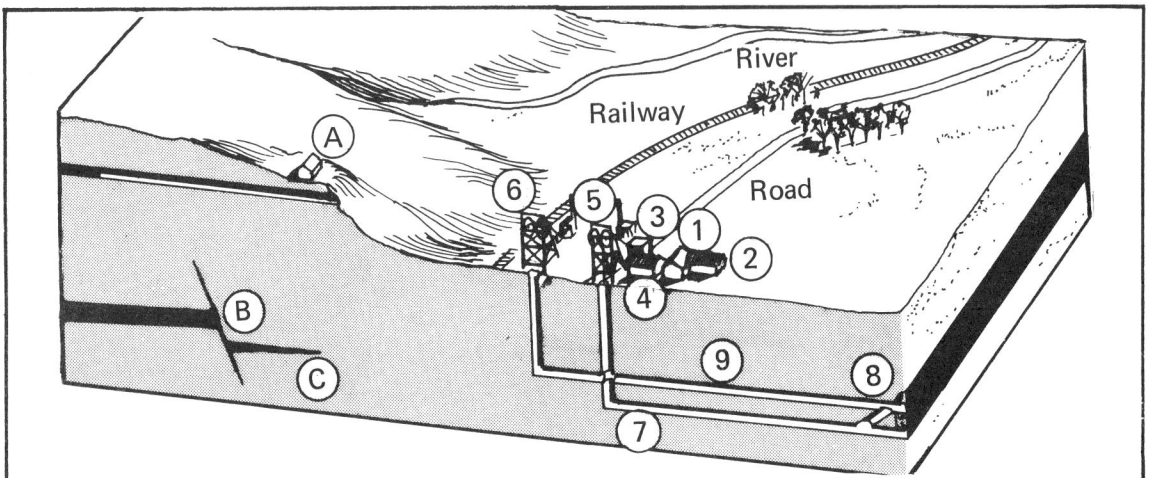

Fig. 21

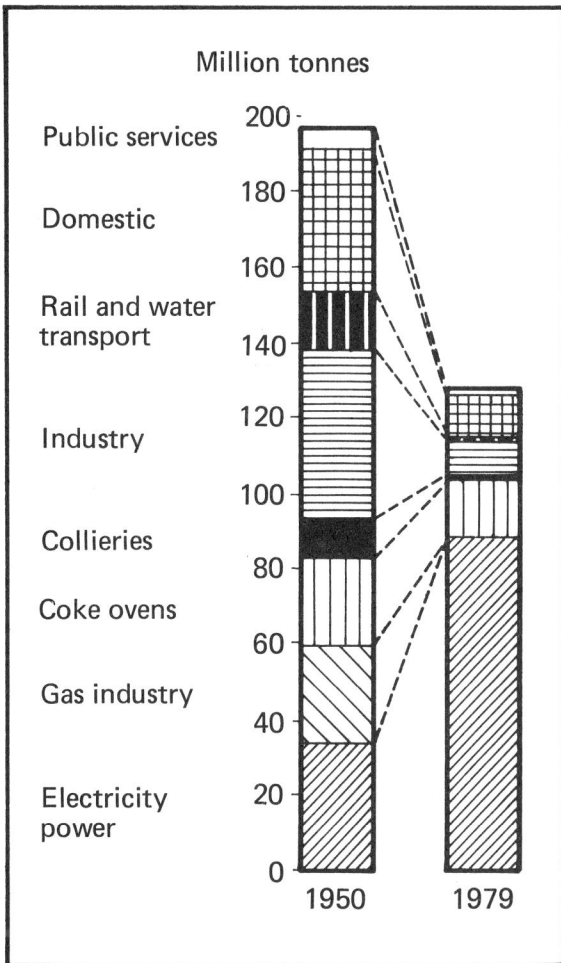

Fig. 22 Use of coal in Britain

During the last thirty years there have been a number of changes in the demand for coal. In Britain new forms of energy – natural gas and oil – have been found. Also we have *imported* oil from other countries. Our need for electricity has gone up and the steel industry is using less *coke* as the works become more efficient. However, many useful products can be made from coal including chemicals, tar and perfumes.

Study Fig. 22. This graph shows the changes in the use of coal from 1950 to 1979.

8   What were the actual total usages in 1950 and 1979?
9   Copy and complete Fig. 23 using the information on this page.
10 Which use has disappeared completely? Why?

The changes you have described are also shown in other aspects of the British coal industry. In particular, many countries are using machinery to cut and transport the coal to the surface. Study Fig. 24 below.

11 Write three sentences describing the changes.
12 How do you account for the increase in output per man?

Use	Increase/decrease	Reason
Public services	Decrease	Change to other fuels
Domestic		

Fig. 23 Use of coal in Britain, 1950-1979

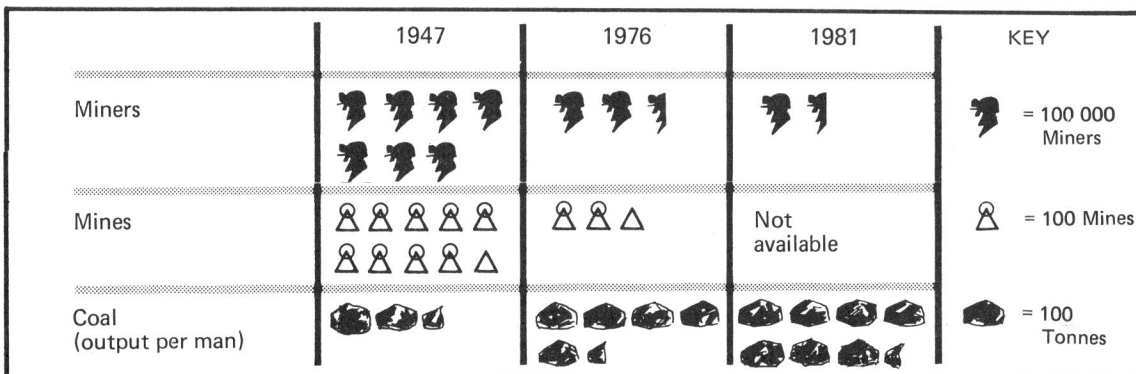

Fig. 24

2B The Ruhr coalfield

The Ruhr coalfield is situated in West Germany and lies to the east of the River Rhine.

1 Make a large copy of Fig. 25 and using an atlas add the following towns: *Duisburg, Essen, Bochum (B), Dortmund, Gelsenkirchen.*

Mining for coal started in the valley of the River Ruhr in the Middle Ages when seams of coal were found near the surface. This is called the *exposed coalfield.* Miners were able to dig the coal by tunnelling into the valley sides to make a *drift mine.* Later, as the demand for coal increased and more coal was found in the region, the mines were extended northwards.

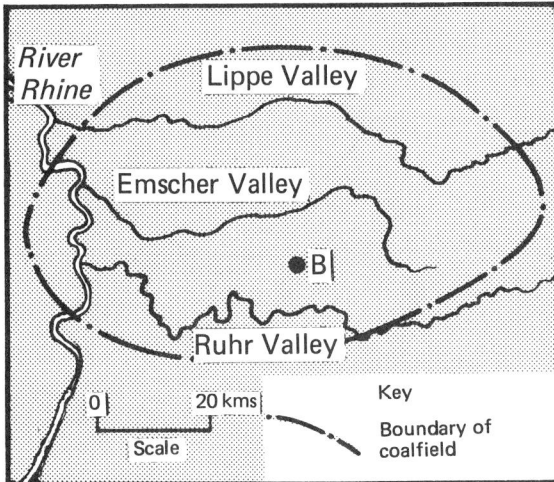

Fig. 27 Population of the Ruhr coalfield

In this direction the coal seams were found to *dip* under other layers of rock, so it became necessary to sink a shaft. The first one was started in 1839. This area of coal is called the *concealed coalfield.* These new mines were larger than the first drift mines which began to decline as the centre of mining shifted northwards from the Ruhr Valley.

As the coalfield grew and coal production increased, the number and size of towns on the coalfield multiplied. Coal became an important raw material for the chemical and iron and steel industries.

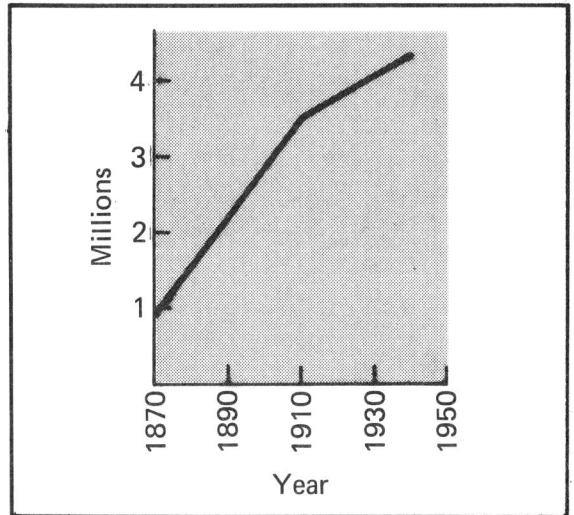

Fig. 25 The Ruhr coalfield

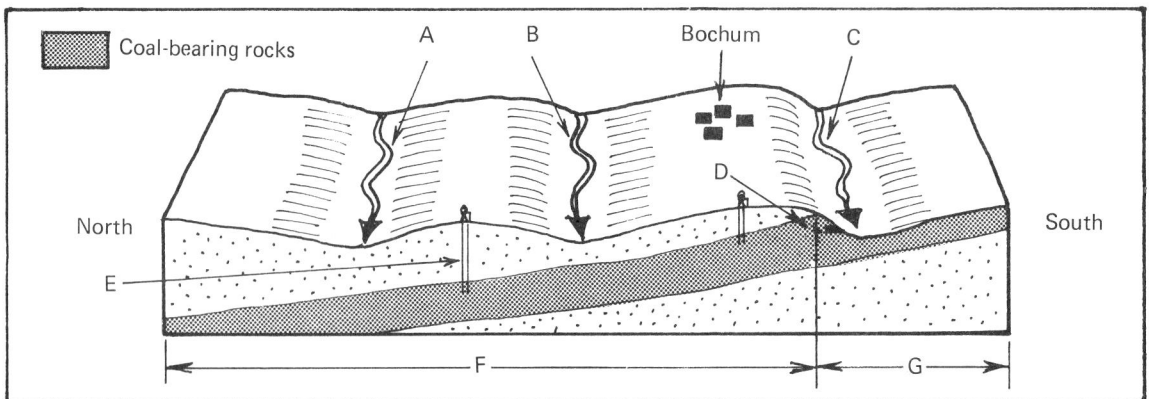

Fig. 26

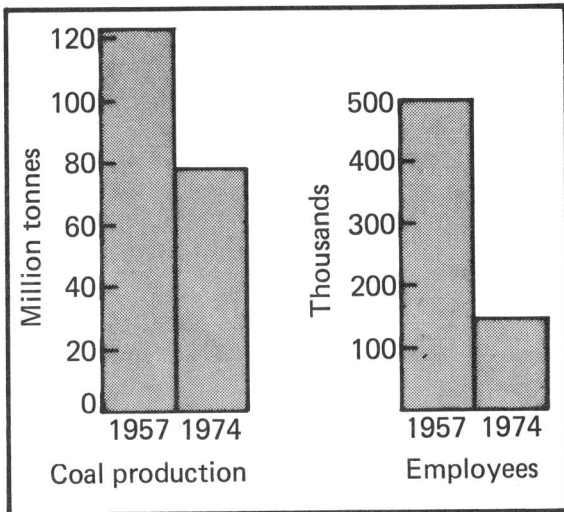

Fig. 28

2   Study the map you have drawn in your exercise book and the block diagram of the Ruhr coalfield, Fig. 26.

Make a copy of the block diagram and add the following names to the correct lettered arrow: *Ruhr Valley, Lippe Valley, Emscher Valley, shaft, drift-mine, exposed coalfield, concealed coalfield.*

3   Study the graph of the Population of the Ruhr Coalfield, Fig. 27. By how many times did the population increase between 1870 and 1930?

In the last thirty years coal-mining has undergone several changes as you saw in the last section on Britain. This is also true of the Ruhr. Study the graph, Fig. 28 which shows figures for coal production and the number of employees in the Ruhr coalfield in 1957 and 1974.

4   What changes have taken place between 1957 and 1974?
5   What are the reasons for these changes?

Bochum is one of several large industrial towns on the Ruhr coalfield which owe their rise to the development of coal mines and later, iron and steel. But the coal seams in and around Bochum are now used up and the industrial character of Bochum is changing.

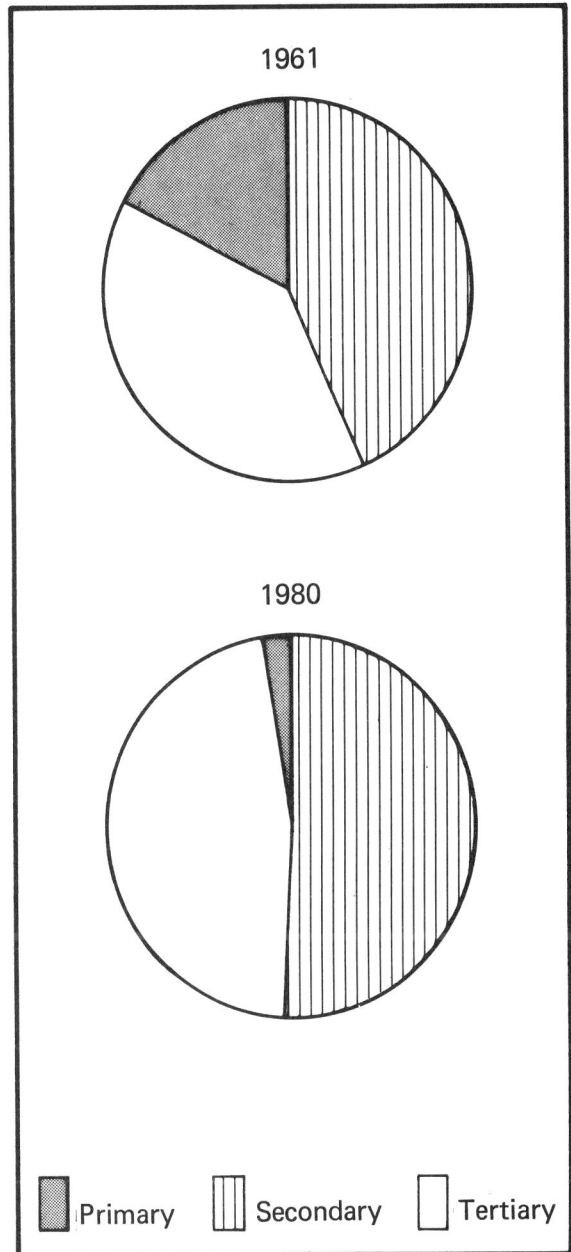

Fig. 29 Employment structure — Bochum

6   Study Fig. 29 which shows the employment structure for Bochum in 1961 and 1980. Describe the changes which have taken place in employment in Bochum.
*7   What effect could mine closures have on towns in the Ruhr?
*8   Using the information on these pages write a brief history of the Ruhr Coalfield.

2C Mining iron ore

Iron ore deposits are found in many parts of the world, particularly in the USSR, Australia, Canada, Venezuela, Sweden and Brazil. These *ores* are rocks containing the metal, iron, in an impure form. The impurities and waste rock are removed from iron ore so that the iron can be used in industry and for making steel. The early stages in this industry are shown in the simple *flow diagram* in Fig. 30.

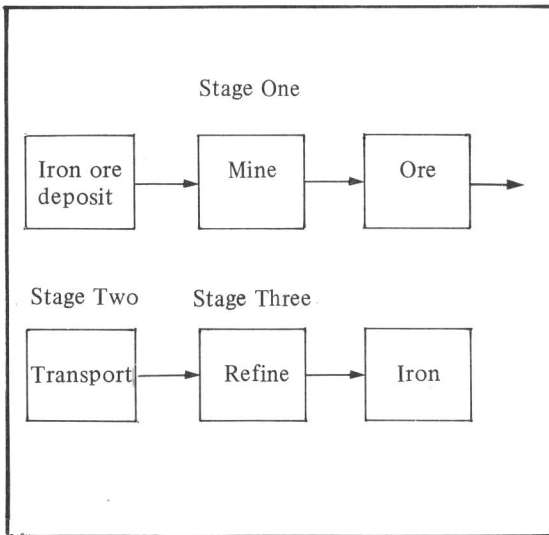

Fig. 30 Iron — from rock to metal

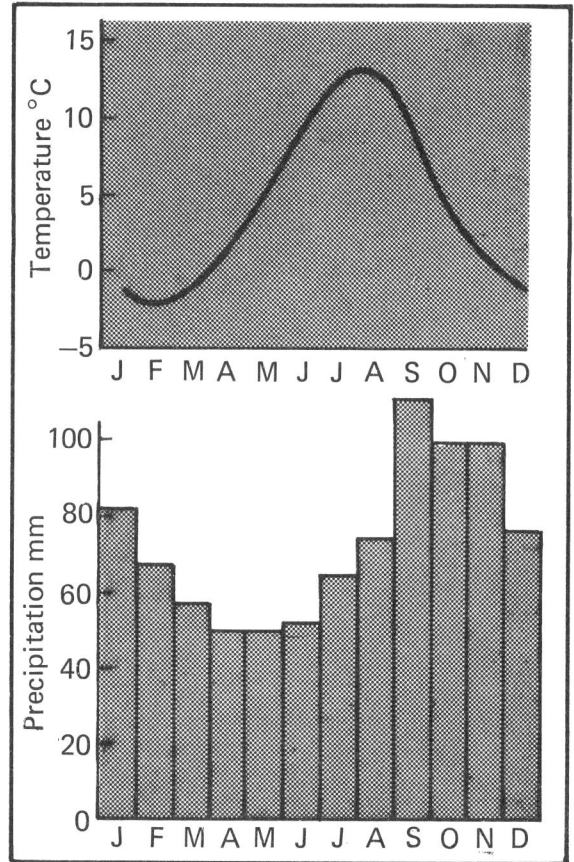

Fig. 32 Average monthly temperature and precipitation figures for Bodo, Norway, 67°N, 12°E

Stage One of this process is called an *extractive* industry because the ore is extracted or removed from the ground. In Sweden, north of the Arctic Circle, lies a large deposit of iron ore which has been mined for about 100 years. The ore deposits are located at Kiruna and Gallivare. Find these places on the map in Fig. 31.

1 Make a copy of Fig. 31 in your exercise book.
2 Using the weather information for Bodo on the graph in Fig. 32, describe the weather conditions for Bodo in winter and summer.
3 How do the January and July figures for Bodo compare with those where you live?

Winter temperatures at Kiruna are about 10°C lower than those at Bodo.

4 Look at the map you have drawn and explain why Bodo on the coast is warmer than Kiruna inland.
5 What is the sea like in the Gulf of Bothnia between December and May?
6 Study the aerial photograph of a winter scene at Kiruna in Fig. 33. Match the letters A, B, C, D to the following: *iron ore mountains (2), frozen lakes, the settlement of Kiruna.*
*7 Study the flow diagram and state which of the stages are linked to the primary, secondary and tertiary sectors of industry.

18

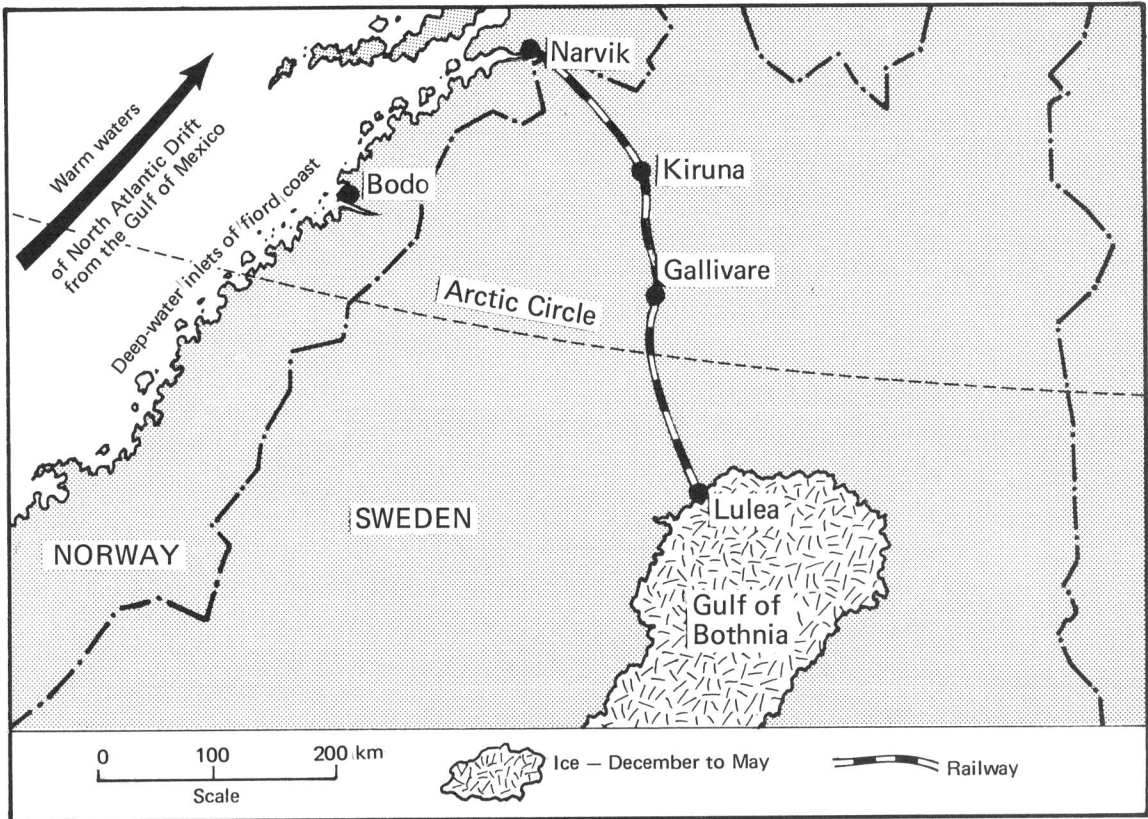

Fig. 31

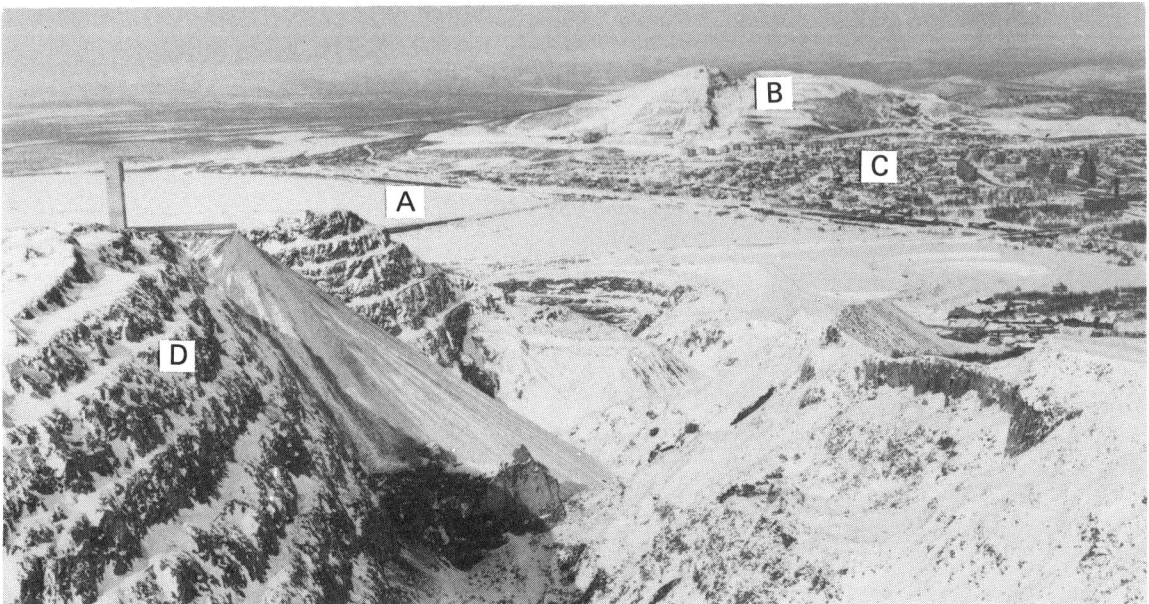

Fig. 33

2D Mining iron ore at Kiruna

Mining at Kiruna used to be the *open-cast* method. This means that it was dug at or near the surface, the ore being removed by mechanical excavators. In Fig. 33 on page 19 you can see the 'step-like' ore mountains that are left as a result of this method of mining. However, now the iron ore deposits left at Kiruna are deep and the mining has changed from surface work to underground work. A part of this underground work can be seen in Fig. 34. A tunnel has been bored into the mountain and the machine is drilling holes into the roof of the tunnel to prepare for blasting. The ore removed by blasting is crushed and taken to the surface. At the surface the iron is separated from the waste rock.

1 Explain the difference between open-cast mining and underground mining.
2 As the surface ores became used up at Kiruna, what change in mining took place?
3 Describe the work that is taking place in Fig. 34.

Fig. 35

Fig. 34

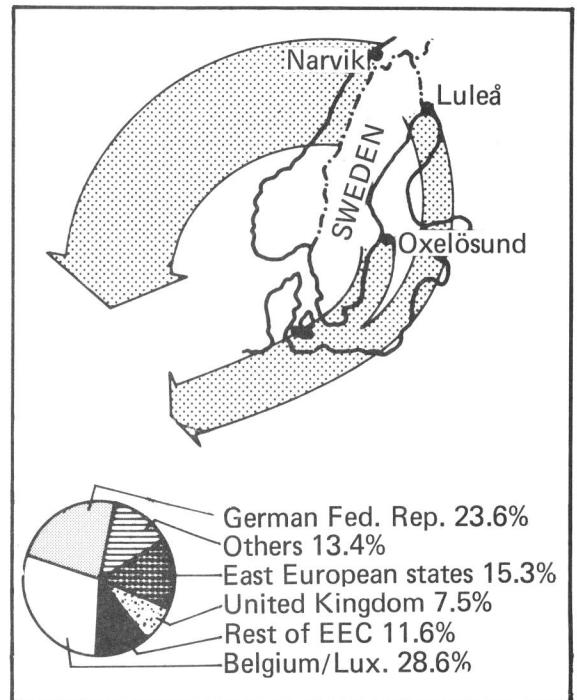

German Fed. Rep. 23.6%
Others 13.4%
East European states 15.3%
United Kingdom 7.5%
Rest of EEC 11.6%
Belgium/Lux. 28.6%

Fig. 37 Swedish iron ore exports in 1978: main shipping ports and importing countries

Fig. 36

When the crushing is complete and the iron ore has been made into medium-sized pellets, it is ready for transport by rail to the coast. Look again at the map on page 19.

4 Which routes can be used to move the iron ore to the coast?
5 How long is each route?
6 Considering the coastal conditions and position of Kiruna, which route is likely to be used most? Why?
7 Look at Fig. 35. This shows an iron ore train on the way to the coast. Describe the country-side it is passing through.

At the port of Narvik the iron ore is loaded on to ore-carrying ships. This operation is shown in Fig. 36. The crane grabs move the iron ore from the *stockyards* (storage areas) to conveyor belts which carry the ore to the ships waiting at the quay in the sheltered, deep-water inlet. The ore is then ***exported,*** that is, sold to many countries.

8 Describe how the iron ore is being loaded in the photograph.
9 Study Fig. 37 which shows the main iron ore importing countries. Why is more ore exported from Narvik than from Lulea?
10 Which countries buy most of Sweden's iron ore?
*11 Draw a flow diagram to show the journey made by the iron ore from the underground mine to an importing country.
*12 Make a simple sketch-drawing of the photograph in Fig. 36. Label the drawing as follows: *iron ore stockyard, crane grab, quay, ore-carrying ship, town of Narvik, mountains.*

2E Quarrying china clay

The mineral china clay or kaolin is found in Britain in Devon and Cornwall. It is a very useful mineral and is used in the manufacture of paper, pottery, rubber, paint, plastics, insecticides and fertilisers. The deposits of china clay are found where there are granite rocks in South-West England. Millions of years ago parts of the granite rock were changed by very hot gases and liquids coming from deep in the earth. Granite consists of three minerals – *quartz, mica* and *feldspar.* The feldspar in the granite was changed to a soft white powder – china clay.

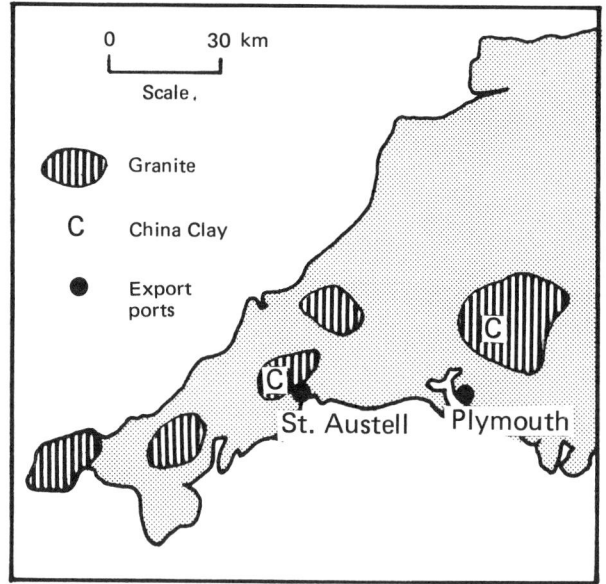

Fig. 40

Fig. 38

Fig. 38 shows china clay being taken from the ground. These are the various processes involved:

A   Remove the ***overburden*** (soil and rock lying on top of the china clay).

B   Blast down granite that has not fully changed to china clay.

C   Wash out china clay using high pressure jets of water known as *monitors*.

D   Pump ***slurry*** (mixture of clay, impurities and water) to the surface.

E   Take the impurities, mostly sand, to surface tip.

1   Fig. 39 is a sketch of the photograph. Copy or trace the sketch and complete it by adding the following labels in the correct places: *old tree-covered tips, tips being landscaped, sand tip, overburden, road to surface, pit workface, dumper truck, monitor, slurry.*

2   Copy Fig. 40 which shows the granite areas of Devon and Cornwall. Using an atlas, name Dartmoor and Land's End.

After removal from the pit the slurry is pumped to the surface and into large tanks. Here it is refined and dried. The china clay is now ready for delivery to many industries both in Britain and abroad.

3   Study the figures in Fig. 41 which show recent china clay production. What proportion of china clay is exported?

4   Study your map and name the two exporting ports.

	1974	1977	1981
Production (000 tonnes)	4 284	4 338	2 629
Exports (000 tonnes)	2 456	2 267	2 196
Value of production (£000)	46 151	73 995	211 918

Fig. 41 UK china clay production

Fig. 39

2F China clay and the environment

One of the problems that faces the china clay industry is how to get rid of the waste it produces. For every ton of clay removed there are six to seven tons of waste. The sand waste is usually piled in large coneshaped tips (see Fig. 38, page 22). Some of the older tips have become covered with vegetation because some soil was mixed with the sand. Because of the danger of landslides, tips today have to be made in layers. These new tips with flat surfaces can be sown with vegetation. Many experiments have been carried out to see how vegetation can best be grown on the tips. As far as possible active working areas are landscaped by planting trees, but it takes a long time and costs a lot of money to make big improvements.

1   Look at the photograph of West Carclaze Pit in Fig. 38 on page 22. Describe the damage that this extractive industry does to the environment.

2   Study Fig. 42 which shows how the china clay industry is trying to improve the environment around its works. Copy the diagram and then, in your own words, describe what happens to 8 tonnes of extracted material under the old and new methods.

3   Read the introduction again. Why are new tips made in layers?

4   How are active working areas landscaped?

Study Fig. 43 and Fig. 44. Fig. 43 shows the White River near St. Austell in Cornwall in 1971. It is carrying mica waste to the sea. Fig. 44 shows the same river in 1973 after the new methods of removing the mica waste had been introduced.

5   Describe and explain the changes that took place on the White River between 1971 and 1973.

6   Why do you think the local people would be pleased about this kind of improvement?

*7 It is proposed to have more china clay workings on Dartmoor. Imagine you are one of the following and write a letter to the local paper giving your views on the development:
    — Managing Director of the china clay firm
    — naturalist living nearby on Dartmoor
    — unemployed man living a few miles away.

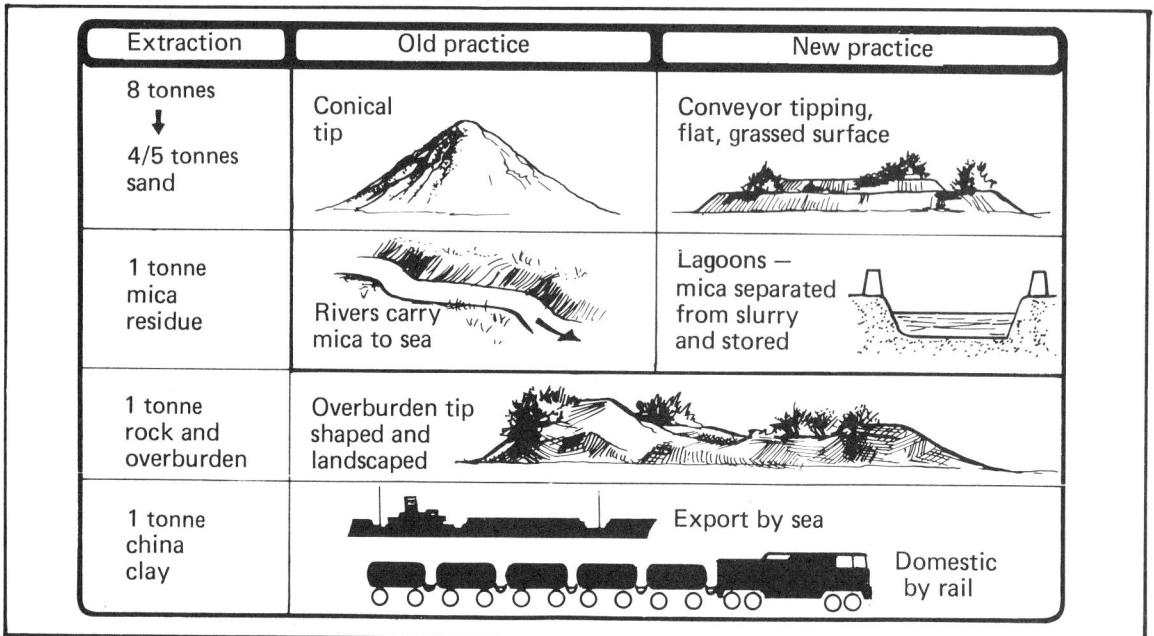

Extraction	Old practice	New practice
8 tonnes ↓ 4/5 tonnes sand	Conical tip	Conveyor tipping, flat, grassed surface
1 tonne mica residue	Rivers carry mica to sea	Lagoons — mica separated from slurry and stored
1 tonne rock and overburden	Overburden tip shaped and landscaped	
1 tonne china clay		Export by sea / Domestic by rail

Fig. 42

Fig. 43

Fig. 44

25

2G Forestry — a primary industry

Wood has been used by Man for thousands of years. It is still one of our most useful materials.

1 Fig. 45 shows some of the present-day uses of wood. Make a list of the uses shown and add as many more as you can think of.

For a number of reasons Britain's woodland has been steadily cleared over the centuries — to make farmland, for boat building, to produce charcoal for iron making, and for building in two World Wars, when we were cut off from many of our overseas supplies.

In order to help protect our forests and plant new ones the *Forestry Commission* was set up in 1919. It is now the country's largest landowner.

Look at Fig. 46 which shows woodland near Windsor Castle. These **hardwood trees** are *deciduous*, that is they lose their leaves in winter. Woodland of this sort once covered much of Britain.

2 What do you notice about the general shape of the trees?
3 Is there a variety of trees, or are they all of the same type?

Fig. 45

Fig. 46

Now look at the photograph in Fig. 47 which was taken on the North York Moors. It shows the types of *coniferous softwood trees* that are typical of the Forestry Commission's forests. These trees are cone-bearing and *evergreen* as they do not lose their needle-like leaves in winter.

Fig. 47

4 What do you notice about the general shape of these trees?
5 Is there a variety of trees, or are they all of the same type?
6 Can you name two things that suggest to you that the forest is man-made and not natural?
7 Make a copy of Fig. 48 which shows eight important trees to be seen in our forests. Complete the information on the diagram.
8 Explain the following terms in your own words: *deciduous, hardwood, evergreen, coniferous, softwood, forest.*
*9 Deciduous and evergreen trees mature at different rates. For foresters what is the main advantage of coniferous trees compared with deciduous trees?

*10 *Project idea*
Use the library to find out more about the Forestry Commission.
*11 Use an atlas to find the world's biggest areas of deciduous and coniferous forests. Draw your own map to show their location.

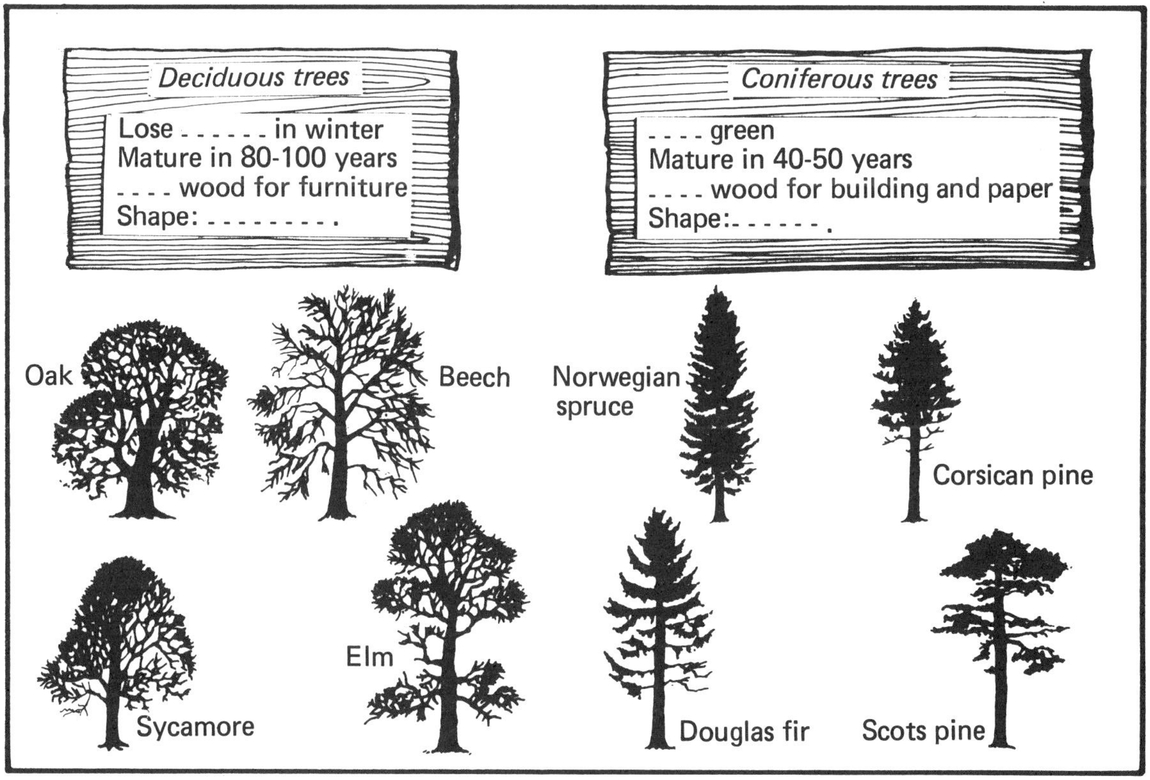

Deciduous trees
Lose - - - - - - in winter
Mature in 80-100 years
- - - - wood for furniture
Shape: - - - - - - - - .

Coniferous trees
- - - - green
Mature in 40-50 years
- - - - wood for building and paper
Shape: - - - - - - .

Oak   Beech   Norwegian spruce   Corsican pine

Sycamore   Elm   Douglas fir   Scots pine

Fig. 48

2H Forestry in Sweden

Fig. 49

Sweden has long understood that her forests, which cover much of the land, are one of her most valuable resources. Fig. 49 shows typically forested country south of Jönköping in Southern Sweden.

1 Find Sweden in your atlas. On which lake is Jönköping situated?
2 Write a paragraph to describe the scene. Mention the following points: shape, slope and height of the land, amount and type of forest, other land use, *density of population.*
3 Fig. 50 shows some information about Sweden's forests. Make a copy of the map and add a key to name the towns, countries and lakes shown by their first letters.
4 What is the main difference between the forests in Southern Sweden and those in Northern Sweden?
5 Why is there no forest in North-West Sweden?
6 Only one hardy, deciduous tree grows in the far north, in the coniferous forest zone. Which tree is this?
7 Look at the bar graph on the map and then say what percentage of Sweden's trees are coniferous.
*8 Make a copy of the bar graph in Fig. 51 and work out the percentage of each type of land use in Sweden.
    Why do you think so much of Sweden is forested?

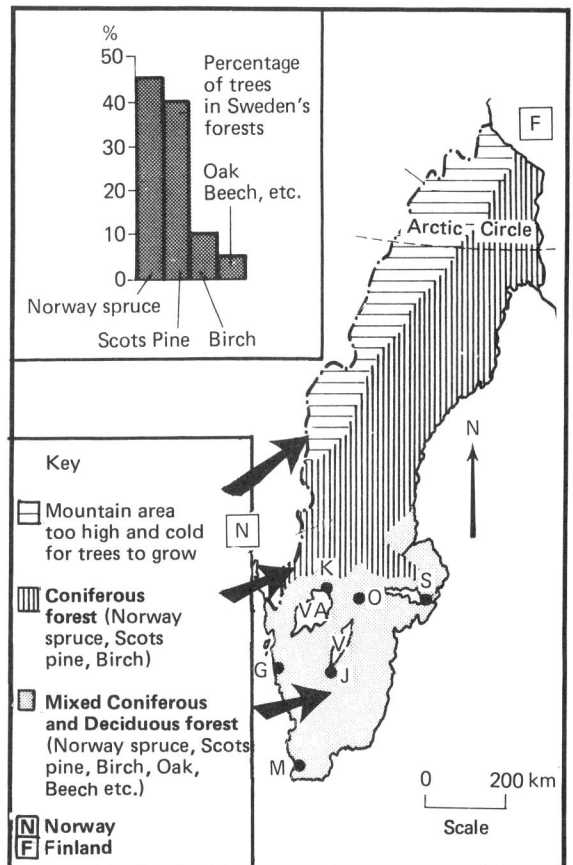

Fig. 50

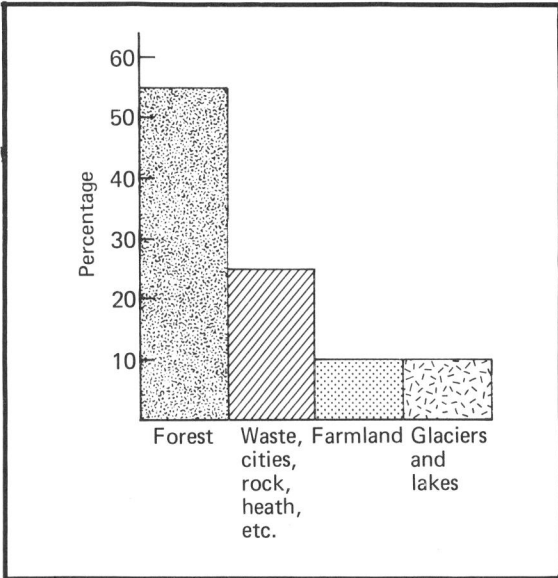

Fig. 51 Land use in Sweden

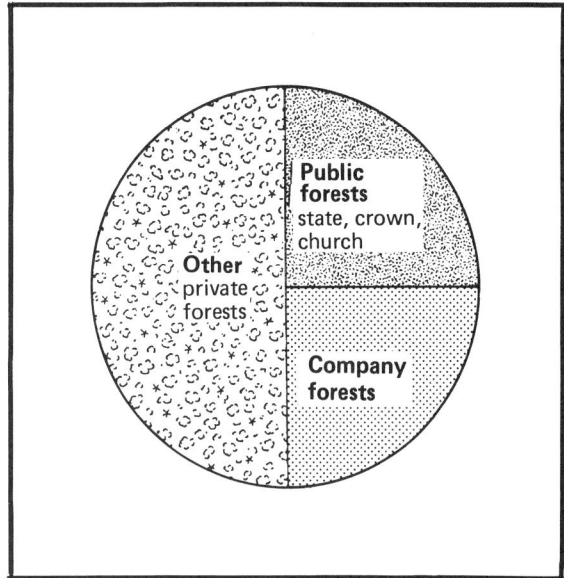

Fig. 53

Fig. 52 shows the values of Sweden's main exports.

9 List the exports in rank order starting with the most valuable.

10 About how many times more valuable are 'Forest Products' than 'Cars and Parts'?

*11 What percentage of total export value is made up by 'Forest Products'?

Fig. 53 shows who owns the forests in Sweden.

12 Write out the following sentences completing them with information given by the pie-graph.

Public forests make up _____ percent of Sweden's forests. This is the same percentage as that for _____ . The remaining _____ percent of the forest area is made up of private forests.

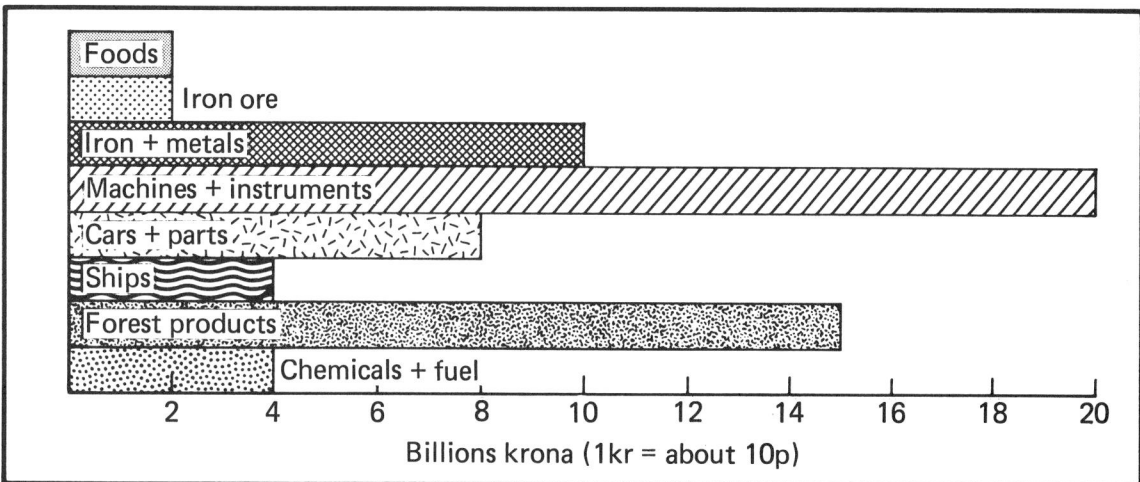

Fig. 52 Sweden's exports

21 Farming the forests

Sweden's forests have been carefully managed for many years. There are documents about forest care going back 400 years. It is because the forests have been so carefully managed and renewed that they are such a valuable money earner for the country. In order to keep producing timber, new forest must be planted to replace the trees which have been cut. Existing forests must be looked after and cared for. Timber is a crop, just like wheat or grass, except that it takes many years to grow to maturity. The process of forest care falls into several stages.

The first thing to do is to *clear* the area of vegetation, then the ground can be ploughed into ridges and furrows by large tractors. The soil is now ready for the young trees to be *planted*. Sweden plants about 350 million trees a year.

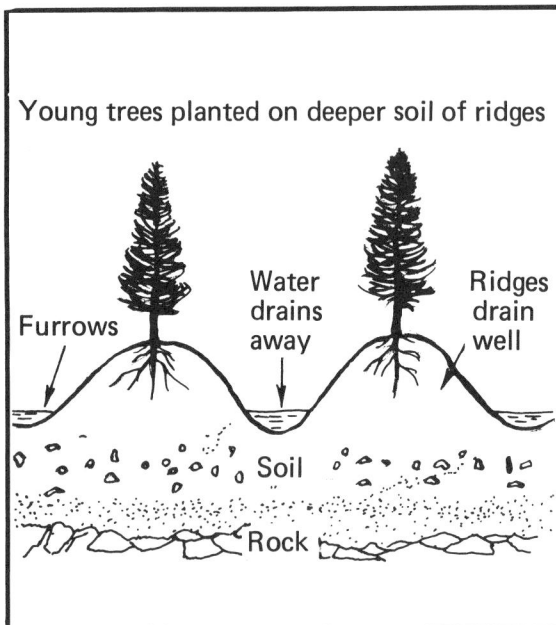

Fig. 55

1 Make a copy of Fig. 54 which shows planted trees. Note two ways in which this method of planting helps tree growth.

Young trees planted on deeper soil of ridges

Furrows

Water drains away

Ridges drain well

Soil

Rock

Fig. 54

Large numbers of healthy, strong, young trees are needed for forest planting. To meet this need all the large forest owners have *nurseries* such as the one shown in Fig. 55. Here young seedlings can start their lives being carefully looked after in ideal conditions.

2 Look at the nursery photograph and name two things which suggest that machinery is used in looking after the trees.

As the forests grow they need *thinning*, so that the best trees have enough room and light.

The last job to be done in the forest is *felling*. Mature trees are cut and prepared for sale and transport.

3 Fig. 56 shows a worker felling a tree with the help of a power saw. Name as many things as you can that are to do with the safety of the worker.

As the cost of labour has risen, more and more forest jobs are being done by machine.

4 Make your own copy of Fig. 57 which sums up the work done in the forests. Complete it by filling in the blank spaces with the correct words. All but one can be taken from the descriptions of forest work.

Fig. 58 shows the European countries which import 75% of Sweden's wood products.

5 Use your atlas to find out which five countries import most wood products from Sweden.
6 Using the scale in the key, work out the percentages of Sweden's wood products exported to each of the five countries.
*7 Forests are also important as areas of *recreation*. Describe how forests can be used in this way.
*8 Recreational use of forests can cause problems for foresters. What problems can result, and how can they be controlled?

Fig. 56

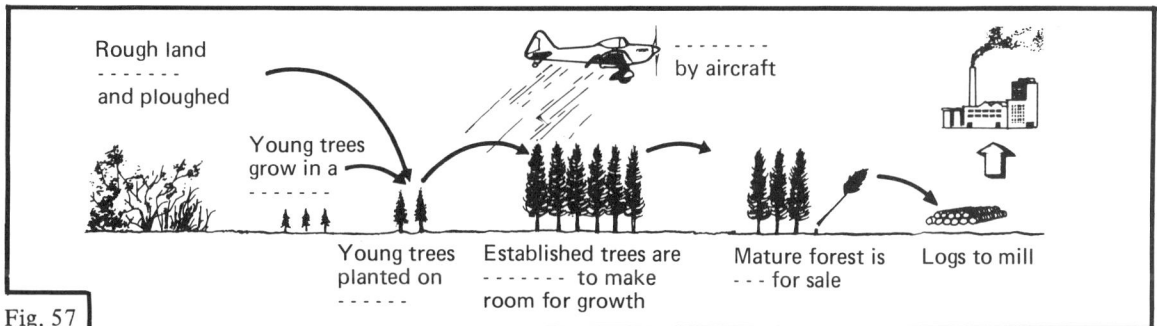

Rough land
- - - - - -
and ploughed

by aircraft

Young trees
grow in a
- - - - - -

Young trees
planted on
- - - - - -

Established trees are
- - - - - - - to make
room for growth

Mature forest is
- - - for sale

Logs to mill

Fig. 57

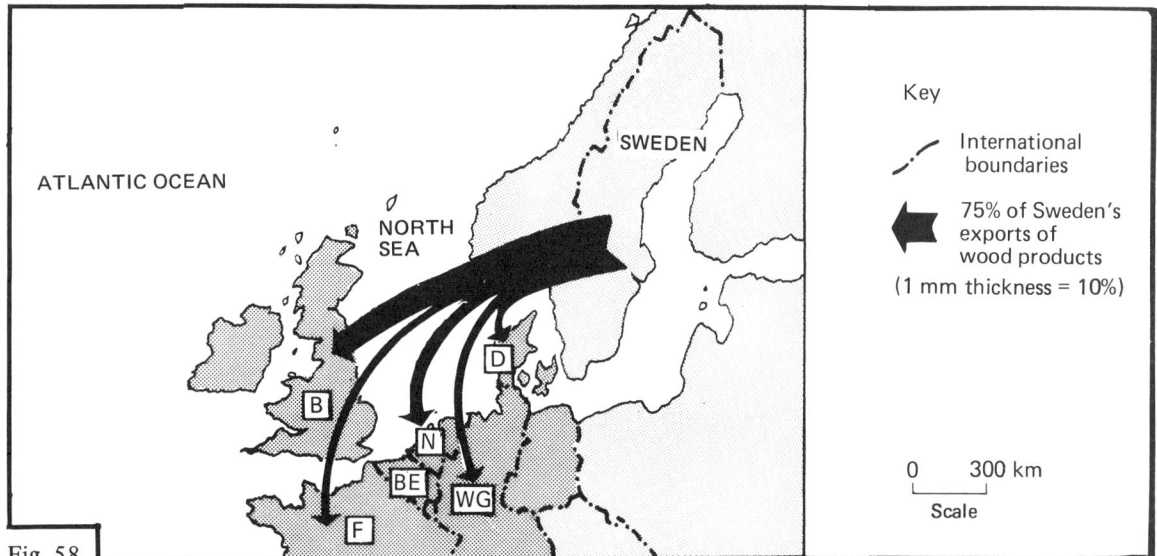

ATLANTIC OCEAN

NORTH SEA

SWEDEN

B

D

N

BE

WG

F

Key

International boundaries

75% of Sweden's exports of wood products
(1 mm thickness = 10%)

0   300 km
Scale

Fig. 58

31

3

Processing the resources

3A Iron and steel — making the steel

The iron and steel industry is one of the most important industries. It makes the materials which are needed in many other industries.

1 Make a list of things found in your home that are made of steel, e.g. refrigerator.

The steel is made from *processing* three basic raw materials – *coal*, *iron ore* and *limestone*. These three raw materials go through various processes or changes before the steel is formed. Fig. 59 shows the major processes in making iron and steel.

2 How are the three basic raw materials extracted from the ground?
3 Which forms of transport are used to move these bulky raw materials?

The three basic raw materials are usually transported to an *integrated iron and steelworks*. This is a large works in which the three basic raw materials are turned first into iron and then into steel. It is known as an integrated works because all the processes take place on one site.

4 Look at Fig. 59 and the two photographs in Fig. 60 and Fig. 61. In your own words describe the main processes that take place in an integrated iron and steelworks from the stockyard to the finished steel.

Fig. 60

Fig. 61

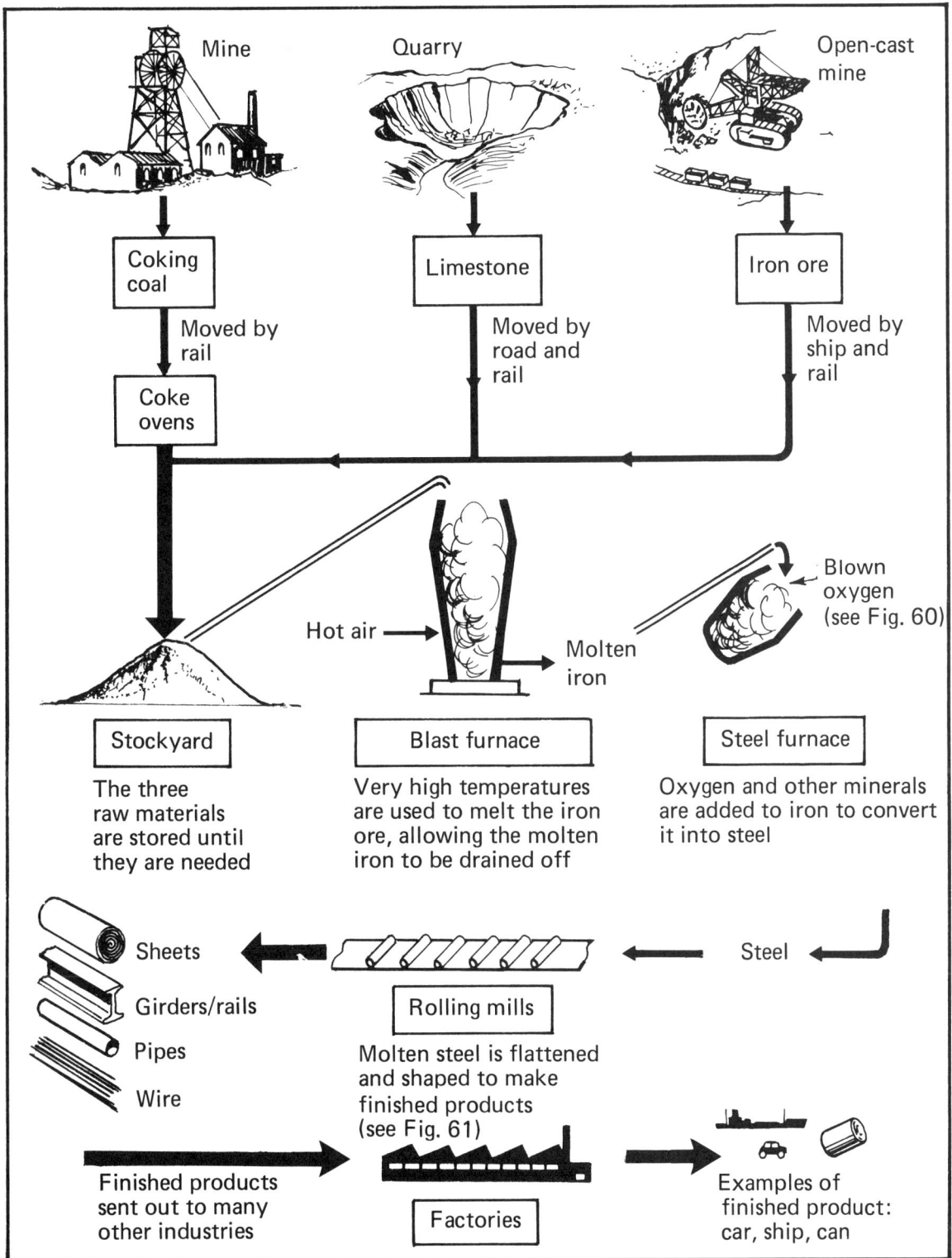

Mine

Quarry

Open-cast mine

Coking coal

Limestone

Iron ore

Moved by rail

Moved by road and rail

Moved by ship and rail

Coke ovens

Hot air →

Molten iron

Blown oxygen (see Fig. 60)

Stockyard

The three raw materials are stored until they are needed

Blast furnace

Very high temperatures are used to melt the iron ore, allowing the molten iron to be drained off

Steel furnace

Oxygen and other minerals are added to iron to convert it into steel

Sheets

Girders/rails

Pipes

Wire

Steel

Rolling mills

Molten steel is flattened and shaped to make finished products (see Fig. 61)

Finished products sent out to many other industries

Factories

Examples of finished product: car, ship, can

Fig. 59

33

3B Iron and steel in South Wales

The South Wales region of the British Isles is one of the major iron and steel-making areas, with two large integrated iron and steelworks.

1 Find the two integrated iron and steelworks on the map in Fig. 62. What are their names?

The iron and steel industry has a long history in South Wales, but over the years there have been major changes in its location. The diagrams in Fig. 63 explain these changes.

2 In what part of the South Wales coalfield were the first iron-making centres located? Name one centre.
3 Where are the two modern integrated iron and steelworks located?

Today all the iron and steel is made near the coast. The cross-section diagram (Fig. 63) explains why.

4 At first all three raw materials were found in the northern part of the coalfield. Which raw materials are no longer found in this area?

5 Where does the iron ore come from that is used in modern steelworks on the coast of South Wales?
6 How do the changes in raw material supply explain the movement of iron and steelworks from the north of the coalfield to the coast?
7 Fig. 64 shows the countries from which Britain imports iron ore. Use your atlas to draw an outline map of the world. Name the countries in the box and draw arrows to show the routes taken by the iron ore ships to Britain.
8 Fig. 65 is an aerial view of the integrated iron and steelworks at Llanwern. Describe the size of the site and the buildings on it.
*9 In your own words describe why Llanwern is a good place for a large steelworks; write about its position in South Wales, and the local relief. Draw a sketch map to illustrate your answer.

Brazil	3 623 009	Norway	942 381
Canada	3 124 104	Mauritania	868 148
Sweden	1 786 360	Venezuela	776 365
Australia	1 744 542	Soviet Union	641 974
S. Africa	1 572 714	Liberia	221 823

Fig. 64 Iron ore imports in 1978 (tonnes)

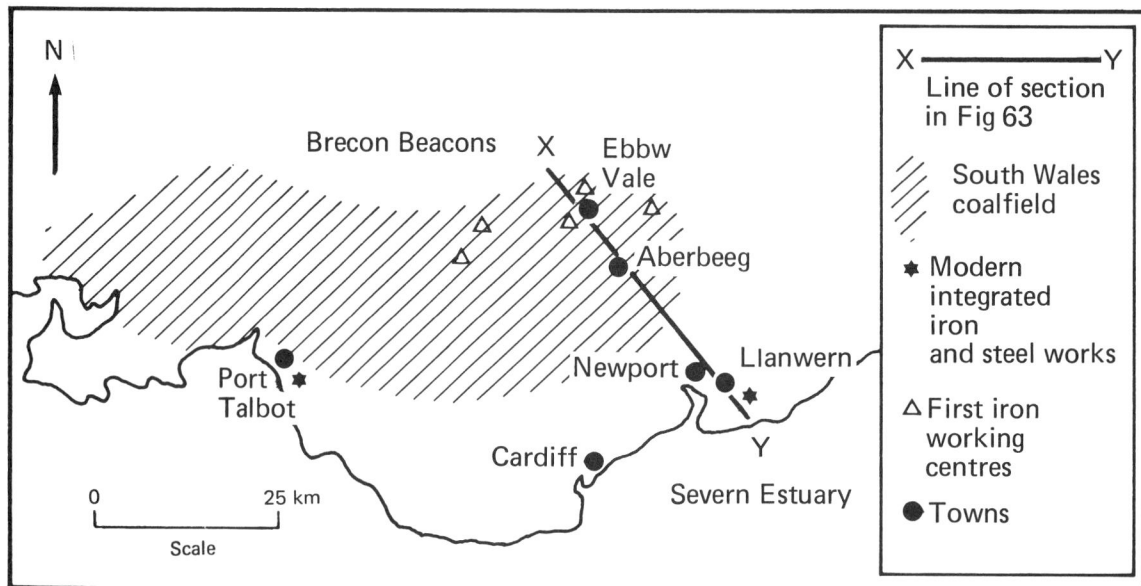

Fig. 62

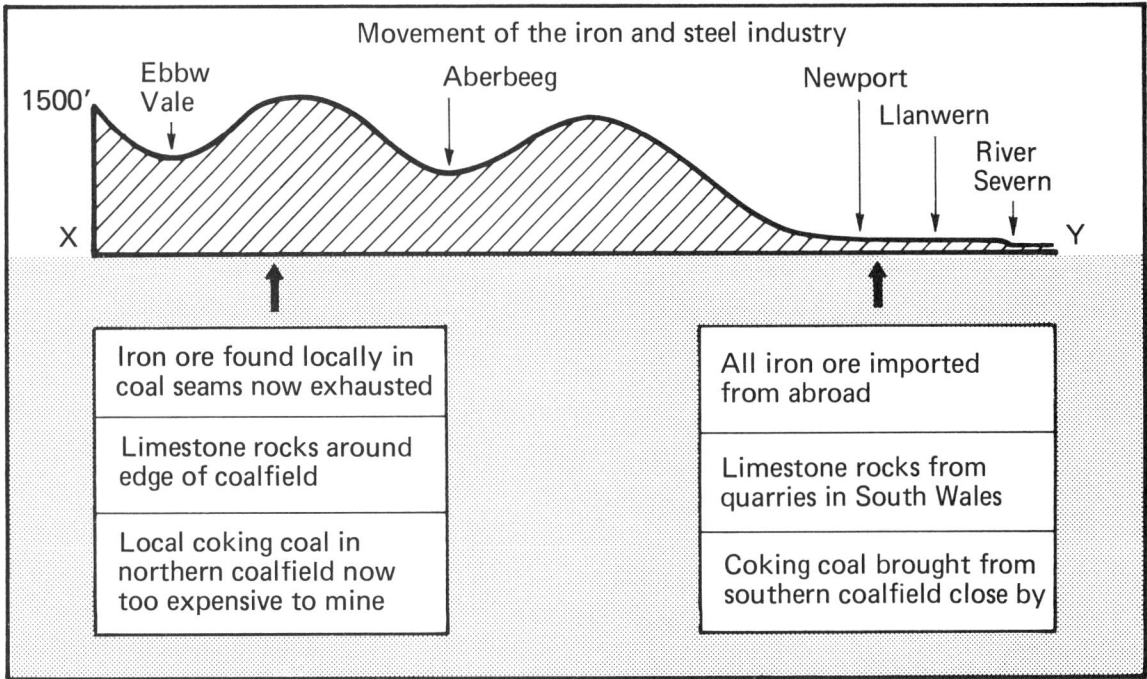

Fig. 63

Fig. 65

3C The iron and steel industry in Britain

There are several regions in Britain producing iron and steel. Fig. 66 shows how much steel was produced in those regions in 1978.

North	2 944
Yorkshire/Humberside	6 625
East Midlands	798
South-East	411
West Midlands	1 492
North-West	196
Wales	6 000
Scotland	1 945

Fig. 66 Regional production in 1978 (thousand tonnes)

1 Study the table. What are the names of the four most important regions for producing steel?
2 Study the map in Fig. 67. For each of the regions you have listed in Question 1 name the major centres of steel production.
3 Using the information on the map write a list of the major steel-making centres, and for each one state which raw materials are found nearby.

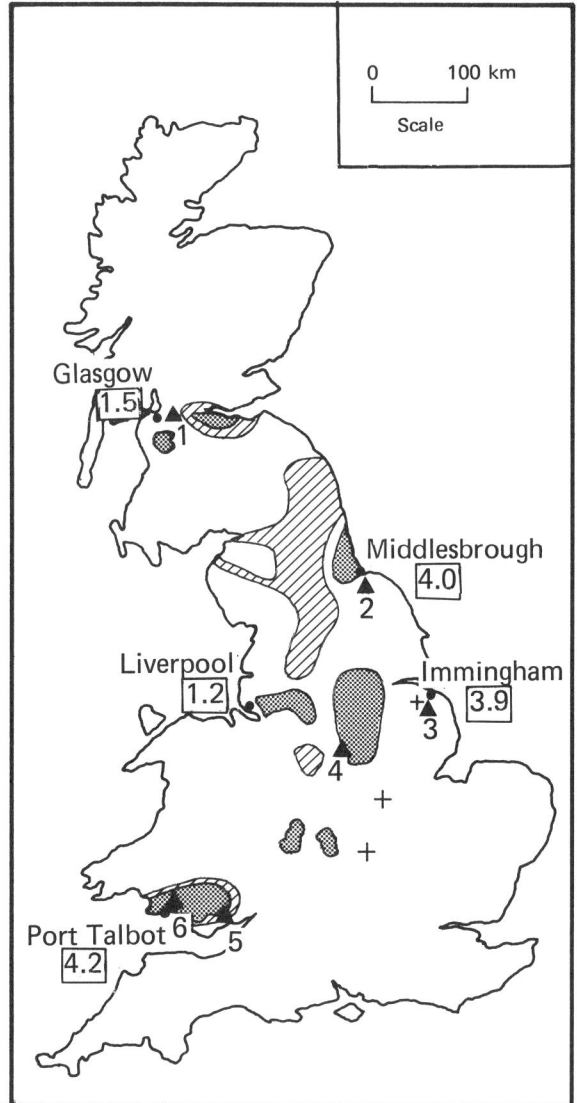

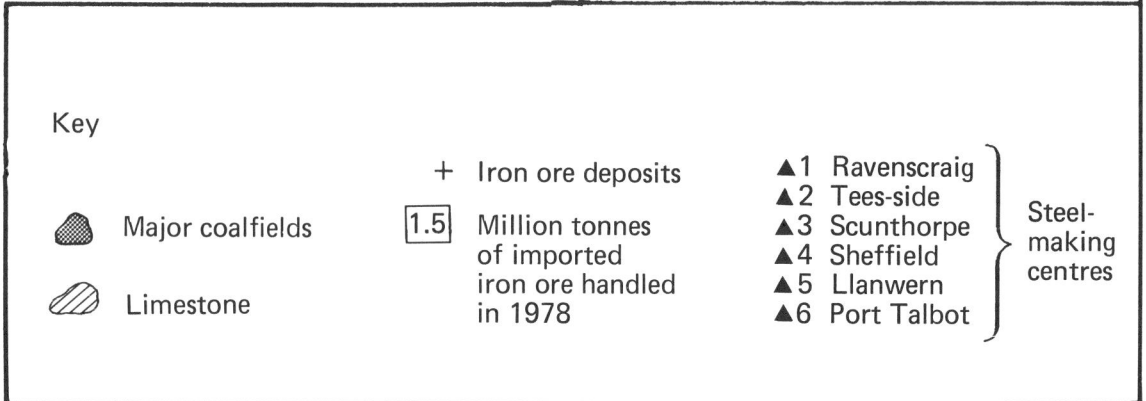

Fig. 67 The iron and steel industry in Britain

Key

⬟ Major coalfields

▨ Limestone

+ Iron ore deposits

1.5 Million tonnes of imported iron ore handled in 1978

▲1 Ravenscraig
▲2 Tees-side
▲3 Scunthorpe
▲4 Sheffield
▲5 Llanwern
▲6 Port Talbot
} Steel-making centres

Much of Britain's iron ore is now imported. The figures for Britain's iron ore imports and domestic production are shown in Fig. 68.

4 Make a copy of the line graph and complete the lines by using the information in Fig. 68. Describe the changes in production and imports that your graph shows.

5 Study the map again. Which ports handle most of the imported iron ore? Match the ports to each steel-producing centre.

6 Study the photograph, Fig. 69. This shows one of these iron ore ports — Port Talbot. Describe the layout of the port and the industrial site in the background.

*7 Select one of the steel-making regions. Draw a sketch map of the region to show the steel-making centre and the sources of its raw materials.

*8 Describe how and why the location of Britain's steel production has changed since the 1800s.

Year	Domestic	Year	Imports
1875	16 067	1913	7 230
1900	14 254	1937	6 950
1925	10 310	1950	8 402
1950	13 171	1958	12 898
1970	12 018	1970	19 923
1975	4 490	1973	21 440
1977	3 745	1977	16 026
1979	4 269	1979	17 924

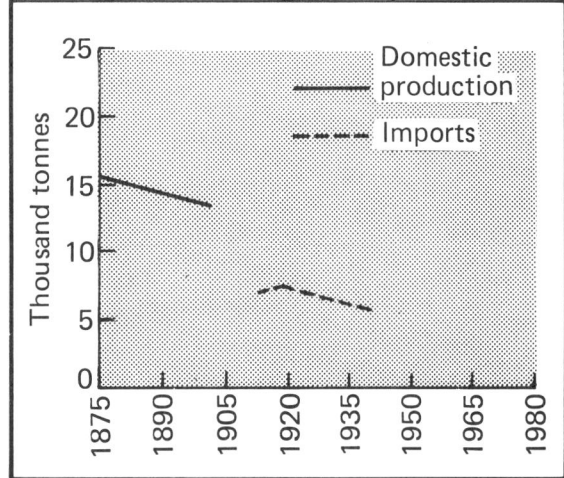

Fig. 68 Britain: iron ore production and imports (thousand tonnes)

Fig. 69

3D Car assembly

The making of cars is one of the biggest industries in the world. It employs many thousands of people and uses large amounts of raw materials. The photographs and information on these two pages are about one company — Ford of Great Britain. There are also other major car manufacturers in the USA, Japan and Western Europe. The parts that are assembled to make a car are called *components,* for example the headlamps. Each component is made from raw materials. The raw materials for the headlamps would be glass and steel.

Fig. 70

1 Draw up a table of car components like the one below, and use the photograph in Fig. 70 to help you fill in each column.

Car component	Raw material(s)
Headlamps	Glass, steel

2 Describe three of the jobs that will be done on a car as it moves along the assembly line.
3 Draw your own diagram of a production line showing four or five stages in the assembly of a car.
*4 What kind of problems may assembly line work have for the men? Suggest how they may be overcome.

The making of cars takes place on an *assembly line*. Each car moves slowly along on a conveyor belt and components are added until the car is completed. The men working on the assembly line repeat the same job on each car many times a day. This means that large numbers of cars can be produced quickly. For example at Dagenham, London, seen in Fig. 71, Ford can make 1250 cars each day. This kind of work is called *mass production*.

Fig. 72 shows a production line in the Dagenham factory.

To make large numbers of cars on an assembly line requires a big factory. This means that a lot of land is needed. For example, the Dagenham Factory in Fig. 71 covers 230 hectares (= 230 football pitches). The factory also requires good transport links. This is necessary to bring in the components and raw materials and to deliver the finished car to the customer.

5 Write out the list below and match the letters to each item in Fig. 71.
   — car engine factory
   — ship unloading raw materials
   — raw materials for blast furnace
   — blast furnace for manufacture of iron
   — finished cars awaiting delivery.
*6 Draw a sketch map of the site and label clearly the main parts of the factory, the River Thames, gas holders and railway lines.

Fig. 71

Fig. 72

Year	Number
1930	3150
1940	10730
1950	16660
1960	55000
1970	67830
1980	73270

Fig. 73 Ford employees in Britain

Factories which assemble cars employ large numbers of men and women. Fig. 73 shows how the Ford Motor Company has increased its workforce.

7 Draw a line graph to represent the figures in the table.
8 How would you explain the steady increase in the number of Ford employees from 1930 to 1980?
*9 Write a few paragraphs to try to explain why Ford built its factory at Dagenham, on this flat riverside land near a large city.

3E Paper making

Much of the world's coniferous softwood is broken down into *pulp*, either by beating or by chemicals. This pulp is the most important raw material in paper making. The making of paper is a secondary industry.

Fig. 74 shows the amount of wood-pulp used in Britain in a year and the countries of its origin. The map is drawn as a *flow diagram.* The quantity of wood-pulp from a particular country is shown by the thickness of the arrow. The scale of thickness is shown in the key.

1 Work out the amount imported from each area and the total.
2 How much wood-pulp is home-produced? What percentage of the total is this?

Fig. 75 shows other products needed in paper making.

3 Make a list of the six things needed to make paper. In your own words say why china clay and size are needed.
4 Fig. 76 shows five possible sites (A—E) for a paper-making factory. Read through the information about paper making on this page and then list the advantages and disadvantages of the five possible sites shown on the map.
5 Which site is most likely to be chosen? Give at least three reasons for your choice.

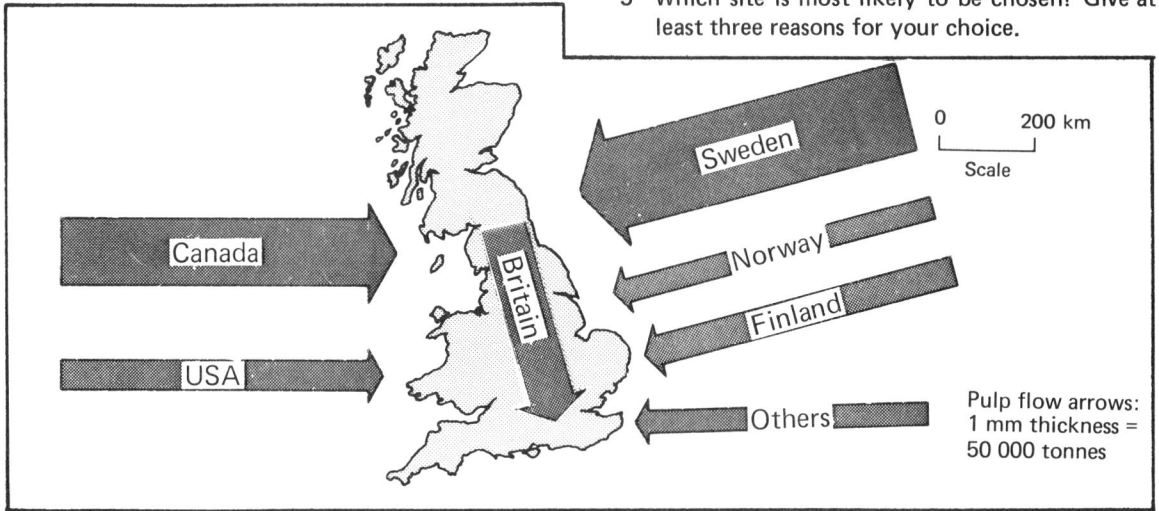

Fig. 74

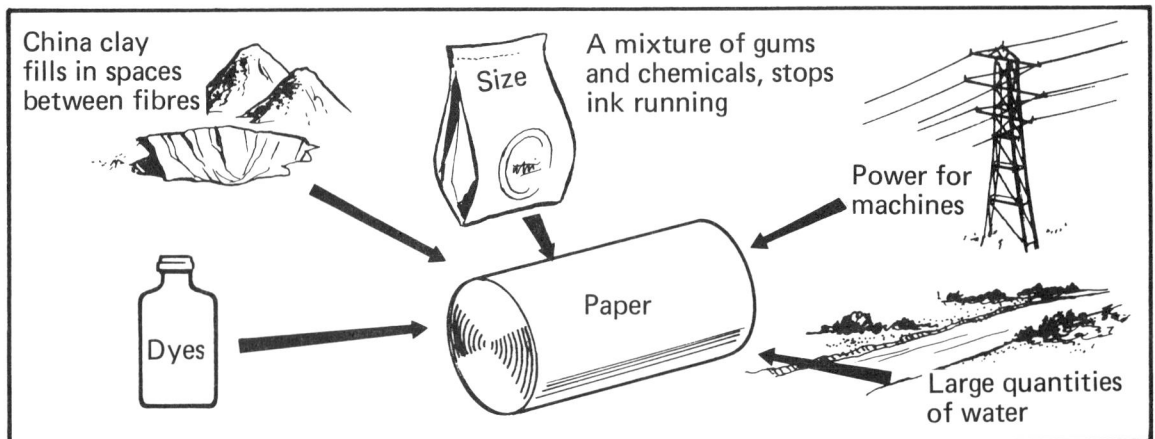

Fig. 75

40

Paper-making involves mixing the pulp and other ingredients with water, letting the fibres settle to form paper and then removing the water. This process needs huge quantities of both water and power. Paper making uses many raw materials, such as oil and coal for power, pulp and china clay that are heavy and costly to transport because they have a low cash value for their weight. The end-product, reeled paper, is also bulky and costly to transport. Paper-making factories are often situated where water transport is available and near a market, that is a large population centre such as a big town.

6 Fig. 77 shows the process of paper making. Make a copy of the diagram and complete it by naming the two outputs shown.
7 Name the four inputs that make up the pulp mixture.
8 In which two ways is water used?
9 Which two inputs are used to make steam?
10 Name one way in which steam is used.

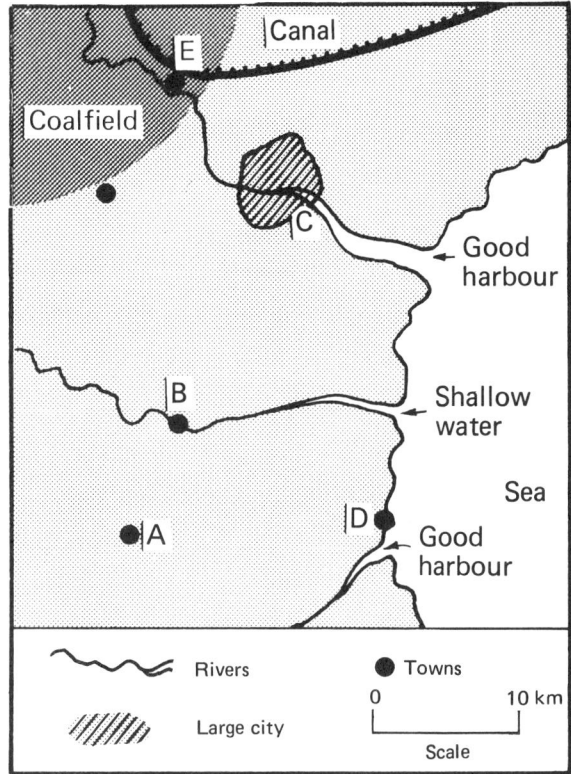

Fig. 76

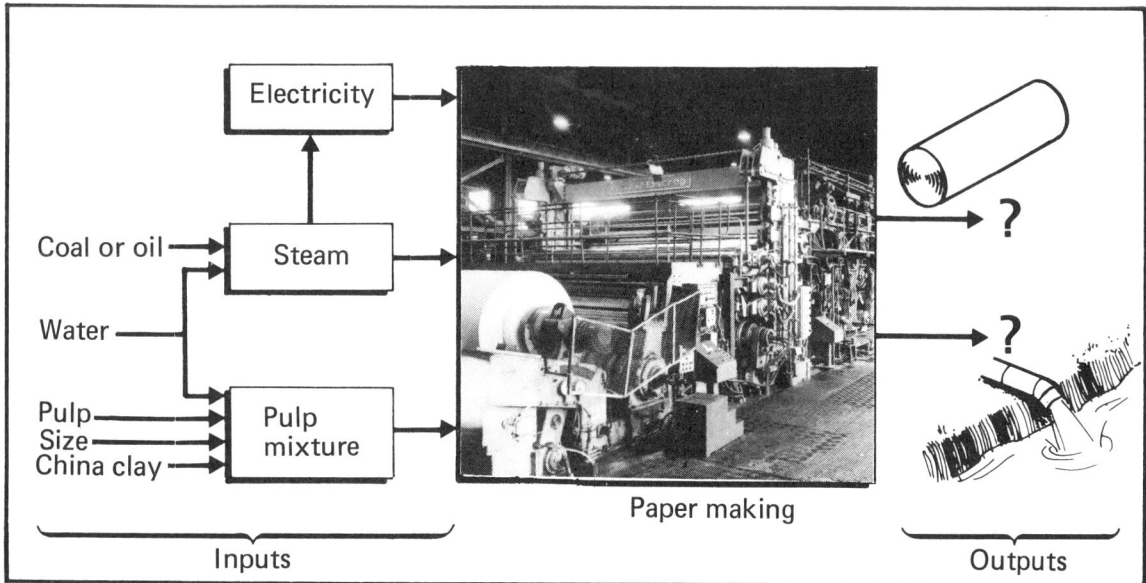

Paper making

Inputs

Outputs

Fig. 77

3F Producing the paper

Wood-pulp arrives at the paper-making factory as a dirty white, dry, compressed bale. Before it can be used to make paper the pulp must be broken down by mixing it with water. This means that all the individual strands, or fibres, become separated. Size and china clay are added to the mixture. The mixture is now ready to start its journey along the paper-making machine.

Fig. 78 shows in diagram form how the wet pulp mixture becomes dry paper.

1   Make a copy of the diagram including the notes.

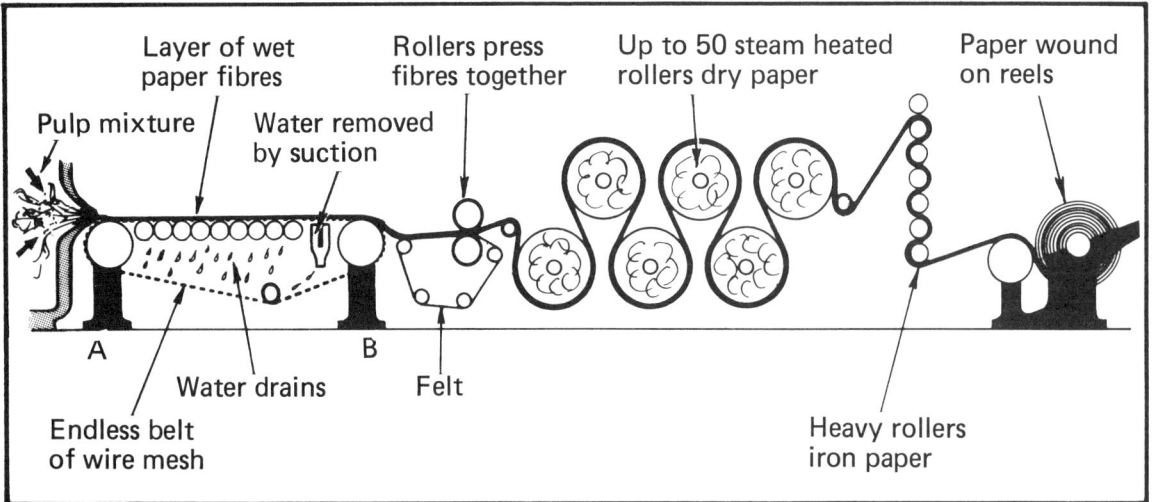

Fig. 78

Fig. 79

42

Fig. 80

Paper-making machines are some of the largest machines in the world. They can be up to 100 metres long. The pulp completes its journey into paper in less than a minute.

Fig. 79 shows a part of a paper-making machine at the start of the process. This is the endless belt that first carries the pulp. It is the section shown between A and B in Fig. 78.

2 Say how long you think this section (A—B) of the paper-making machine shown in the photograph is. Use the height of a man as a scale of two metres.

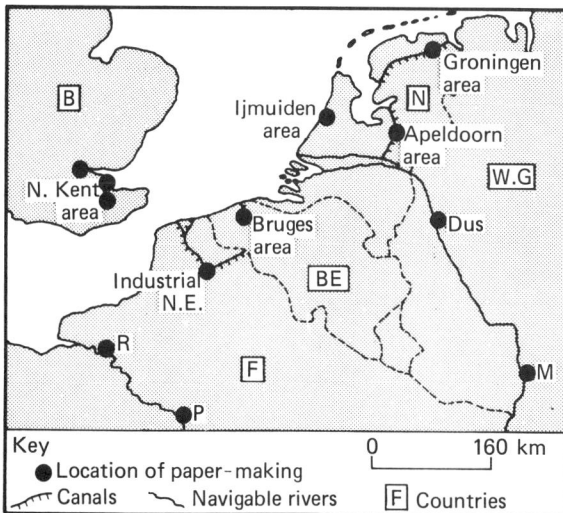

Key
● Location of paper-making
⌒ Canals   ⌒ Navigable rivers   F Countries

Fig. 81

Fig. 80 shows some of the many everyday uses of paper.

3 List all the different uses of paper you can see in the sketch.
*4 Using all the written and diagramatic information given on this double page, write an essay describing the paper-making process from start to finish.
*5 Project idea
Collect as many different types of paper as you can to show its varied uses.

The map in Fig. 81 shows the location of the paper-making industry in part of Europe.

6 Make a copy of the map. Use a key and add the names of the towns and countries shown by their first letters.
7 Choose from the list below and add to your map the symbols that are right for each paper-making location shown.
   — ⌒ coastal site
   — ⌒ good water links to coast
   — ⚘ near large population centre
8 How many paper-making centres are shown on the map?
9 How many centres are on the coast?
10 Name the inland centres.
*11 Write a short essay describing the location of the paper-making industry as shown on the map and explain in your own words some of the important factors that affect the location you describe.

3G Newspapers

The familiar photograph in Fig. 82 shows the last link in the long chain of production and transport that started in the forests of Sweden. Of the paper produced in Britain 7.5% is made into newspapers. This type of paper is called *newsprint*.

1 In which part of industry, primary, secondary or tertiary, is the paper boy shown in Fig. 82 working? Explain your answer.

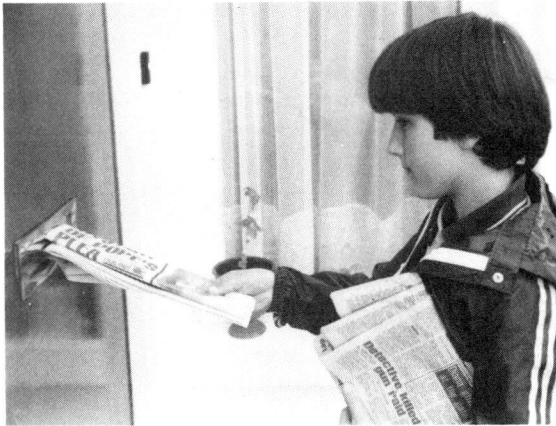

Fig. 82

A study of Fig. 83 which shows the use of newsprint in some European countries, should give you some idea of the large quantities involved. The bar graph is arranged so that the quantities shown are the number of kilos of newsprint used each year per head of population. They are in rank order.

2 How much newsprint do the following countries use per head of population: Sweden, Britain and Germany?
3 Make a list of the countries in rank order according to how many newspapers are sold each day.
*4 Use the population figures given in brackets for each country to work out the total quantity of newsprint used by each country.
*5 Rank the countries according to the total quantities of newsprint used and draw a bar graph to show the information.
*6 Compare the three rank orders. What do you notice? Which two rank orders are very much alike? How do you explain what you have described?

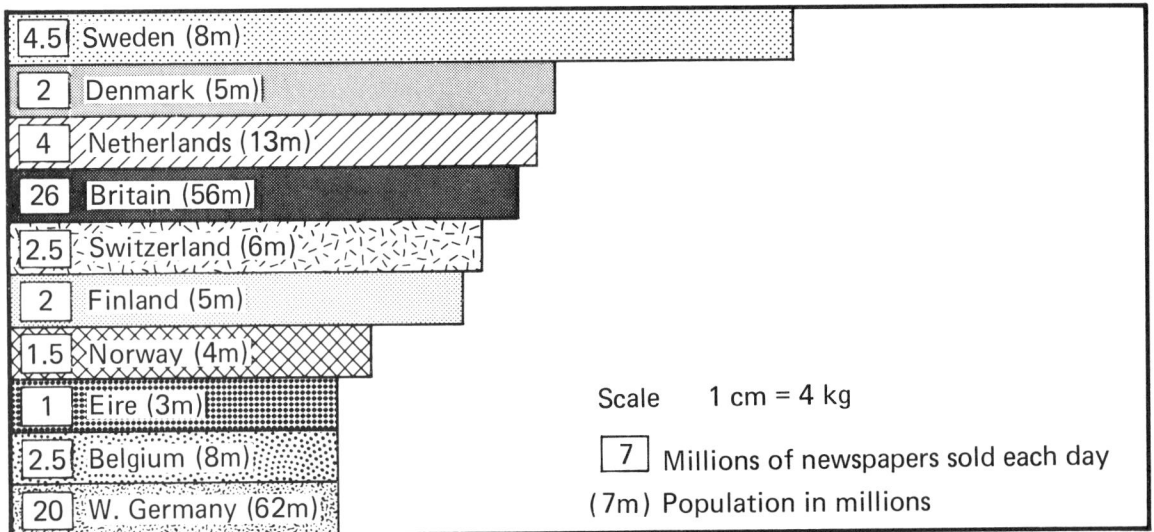

4.5	Sweden (8m)
2	Denmark (5m)
4	Netherlands (13m)
26	Britain (56m)
2.5	Switzerland (6m)
2	Finland (5m)
1.5	Norway (4m)
1	Eire (3m)
2.5	Belgium (8m)
20	W. Germany (62m)

Scale   1 cm = 4 kg

7 Millions of newspapers sold each day

(7m) Population in millions

Fig. 83 Use of newsprint (kg per person each year)

Fig. 84 shows the number sold each day of some European newspapers.

7   Which paper sells most copies? In which country is it published?
8   List the newspapers which sell over 1 million copies each day. How many of these are published in Britain?
9   How many countries have newspapers which sell 1 million or more copies?
10 How do you explain that so few countries have newspapers with very large sales?

The photograph in Fig. 85 shows Fleet Street in London. It is the home of Britain's newspaper industry.

11   Name any newspaper offices that you can see.
*12 Why are so many newspapers printed in London?
*13 Is it a good idea to have this industry in the busy centre of a city with a population of over 10 million? Write a paragraph setting out the arguments for and against the idea.

W. Germany		Sweden	
4 700	Bild-Zeitung	600	Extressen
700	Westdeutsche-Allgemeine		
Britain		**Netherlands**	
3 800	Sun	500	De Telegraaf
3 600	Daily Mirror		
2 300	Daily Express		
1 500	Daily Telegraph		
France			
1 000	France Soir		
1 000	Le Parisien Libere		
800	Ouest France		
Austria			
700	Unabhängige Kronen-Zeitung		

Fig. 84 European newspaper sales (thousands per day)

Fig. 85

45

3H Printing the daily

Fleet Street is the home of many of our famous newspapers.

The making of a newspaper means hurried work, for there is always a rush to get it out on time. Work goes on all through the day and night to get the paper to us by breakfast time. Some idea of the quantities of paper needed can be understood by knowing that it takes 5500 trees to provide the newsprint for one day's issue of one of our popular newspapers.

News, in words and pictures, comes into newspaper offices from all over the world — from reporters, from independent correspondents and from news agencies such as United Press.

Fig. 87 A printing press

Fig. 86

All this information comes into the *Big Room,* shown in Fig. 86. Here the editors and their staff sort the information and shape the stories and headlines into pages for the paper. Pictures are added and facts checked in the reference library.

1 Describe the work taking place in the Big Room.

This prepared material or *copy* goes to the *Composing Room.* Here the words and pictures of each page are made up into metal blocks.

A copy of each page is now made in **papier mché.** This is bent and hollowed into a half-circle and molten metal is forced against it to make the actual metal plate to be fitted to the rollers of the printing press.

The paper can now go to press. The plates are fixed and the reels of paper are positioned. The machine is filled with ink. At full speed a press can print about 800 copies a minute.

2 Describe what happens to the copy when it leaves the Composing Room. Look at Fig. 87 to help you.

As the newspapers come off the press they are automatically folded, sorted and bundled. In the *Publishing Department* the papers are made up into parcels and loaded on to vans and lorries for distribution.

This method of producing newspapers is still common in Fleet Street today. *The Sun* and the *Daily Express* are examples. Some newspapers, however, such as the *Daily Mirror* and *The Times*, use more modern methods. Computers are being used in their Composing Rooms, as you can see in Fig. 88.

Fig. 88

3 Make a copy of Fig. 89 which shows the nightly process of newspaper production. Complete it by adding how the papers are delivered to our homes to the arrow that has the question mark against it.

4 In your own words say what you understand by the following terms: correspondent, Big Room, copy, Composing Room, papier maché, Publishing Department.

*5 Use the diagram to write an essay about the inputs, outputs, and distribution involved in a night's newspaper production.

*6 *Project idea*
Using the library find out what you can about the history of printing.

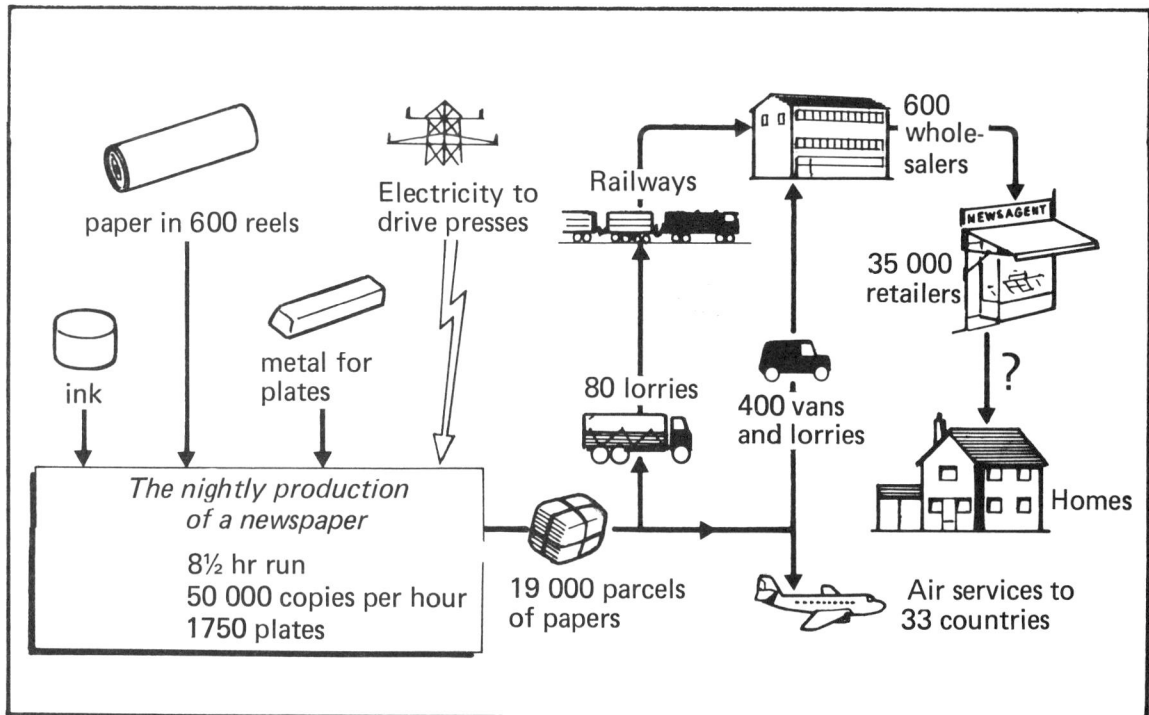

Fig. 89

4

The location of industry

4A Frozen peas — a food processing industry

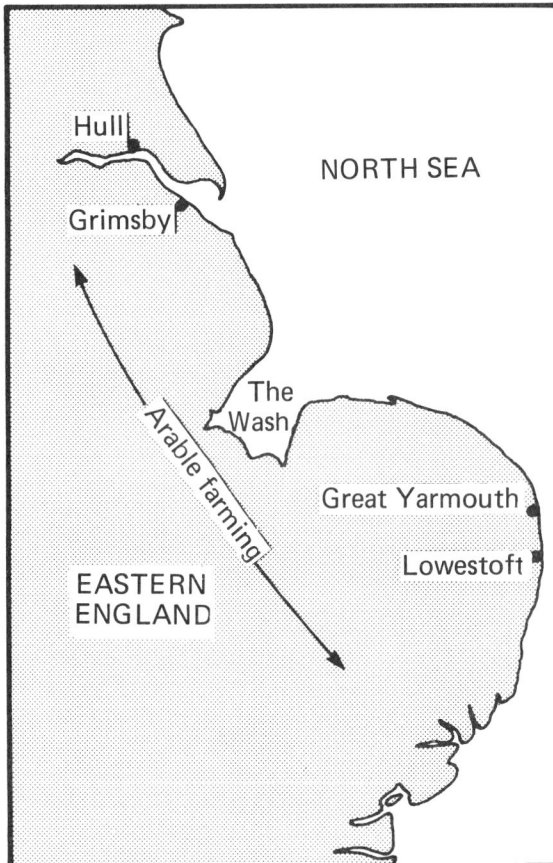

Fig. 90

In March, 1626, Francis Bacon, a courtier and statesman during the reigns of Elizabeth I and James I, stopped his carriage in a London street. He got down from his coach, bought a dead chicken and filled it with snow to see if it would stop it going bad. Unfortunately, as a result of this experiment he caught a chill and died from bronchitis! Three hundred years were to pass before Clarence Birdseye, an American, was to develop a way of freezing foods and selling them in packets.

In Britain a number of vegetables are grown that are especially suitable for *quick-freezing*.

1 Make a list of quick-frozen vegetables that you could buy in your local supermarket.
2 The map in Fig. 90 shows the location of the main freezing centres in eastern England. Use an atlas to name the counties in which each centre is situated.
3 Each centre is a port. Name another important food that could be frozen at these ports.

The surrounding region is an important *arable* farming area of the United Kingdom. This means that it grows a wide range of crops — wheat, barley, sugarbeet, vegetables. In recent years the production of vegetables in this area has increased to meet the growing demand for frozen foods.

Year	Acres	Tonnes
1965/66	131 000	240 000
1970/71	206 000	317 000
1975/76	215 000	359 000

Fig. 91 Production of peas

Period	Tasks
Autumn	Plan for next year's crop. Consider acreage, equipment and seed required.
February to May	Planting. This is spread out so that there is a harvesting season of about 45 days and to spread the work at the processing centres.
June	Samples taken from the fields for testing.
July to August	Harvesting. (Look at Fig. 93 showing a mobile viner; these take up the cut pea vines and thrash out the peas.) This is a fast operation as every pea has to be picked and frozen in less than 2½ hours. Peas are graded, cleaned, quick-frozen and packed in the factory. Look at Fig. 94 which shows millions of peas being carried along a cushion of icy air through the freezing tunnel.

Fig. 92 Annual production cycle for peas

4 Study Fig. 91. This shows the acreage of land used for peas and the weight produced. How many acres were used for growing peas in 1965/66?

5 How much was the acreage in 1975/76? By how much had the acreage increased?

Fig. 93 A pea viner

6 Study Fig. 92 which describes the yearly production cycle for growing and processing peas. Draw a circle and divide it into three segments. In each segment write the tasks carried out in autumn, spring and summer. Give your diagram a title.

7 Why is planting spread out over 45 days?

8 What job is carried out by the mobile viner as seen in Fig. 93?

9 Describe the operations carried out in the factory. Use Fig. 92 to help you.

*10 Explain why it is necessary for farms to be close to a processing factory.

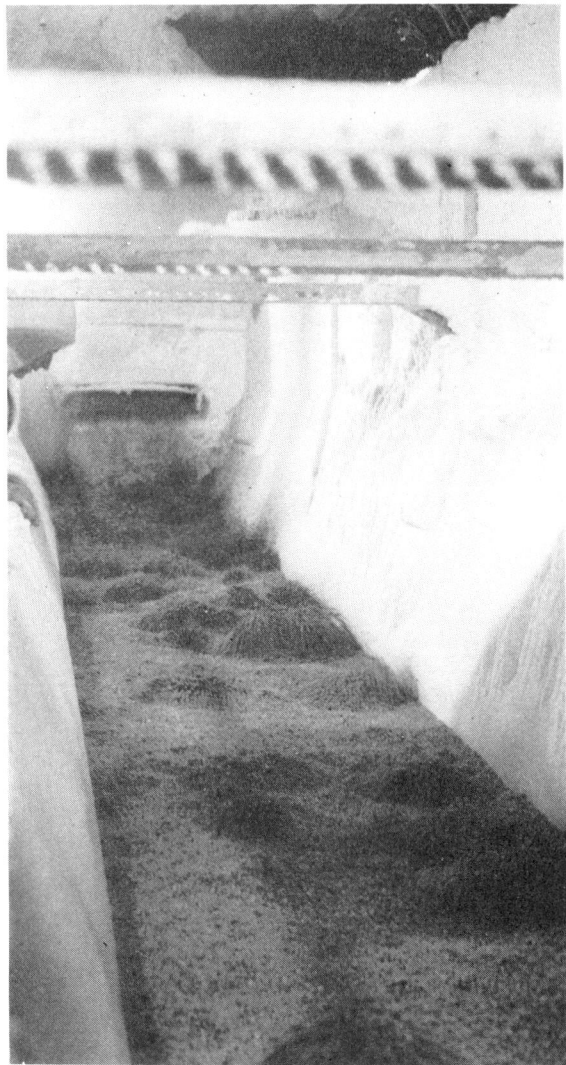

Fig. 94

4B Industry at the coast

Many raw materials needed by industry in a country are imported. This means they are bought from another country. The raw materials are usually transported by sea, particularly if they are *bulky*. A bulky raw material – iron ore, coal, oil – requires a large ship for transport.

1 Study the diagram in Fig. 95 which shows two possible industrial sites for processing imported bulky raw materials. Consider the advantages and disadvantages of each site by copying out and filling in the grid below. Which site do you consider the more suitable?

	Transport costs	Workforce	Market to sell product
A			
B			

In recent years many large industrial sites have been developed at or near coasts. They are known as *break-of-bulk* points because the raw material has to break its journey when the ship reaches the coast. Oil refining, manufacture of iron and steel and petro-chemicals are examples of such industries.

An example of an industrial site located at a break-of-bulk point is shown in Fig. 97. This is the deep-water port of Fos on the south coast of France. Work started on this new port in 1960 as an extension of the port of Marseilles to the east.

2 Using Fig. 96 and your atlas find the position of Marseilles and the mouth of the River Rhône. Describe the location of the ports of Fos and Marseilles.

3 Using the information in Fig. 97, copy out and complete the *flow diagram* in Fig. 98 to show the main features of the iron and steel industry and petro-chemicals industry on the Gulf of Fos.

4 Why do you think it was necessary to dredge a channel in the Gulf of Fos?
Consider the size of supertankers.

*5 With the help of Fig. 99 write a paragraph explaining why Fos was a good position for industrial and port developments.

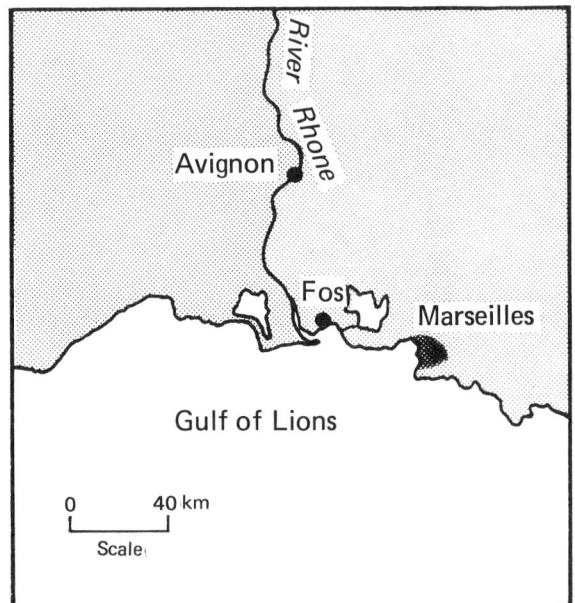

Fig. 96

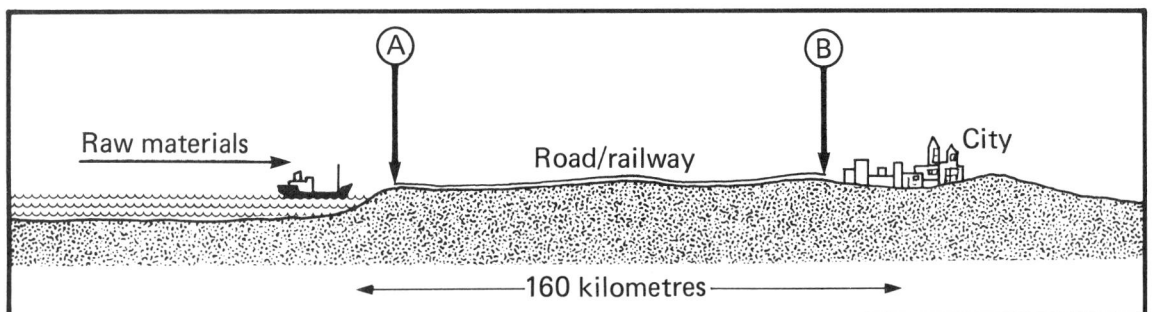

Fig. 95

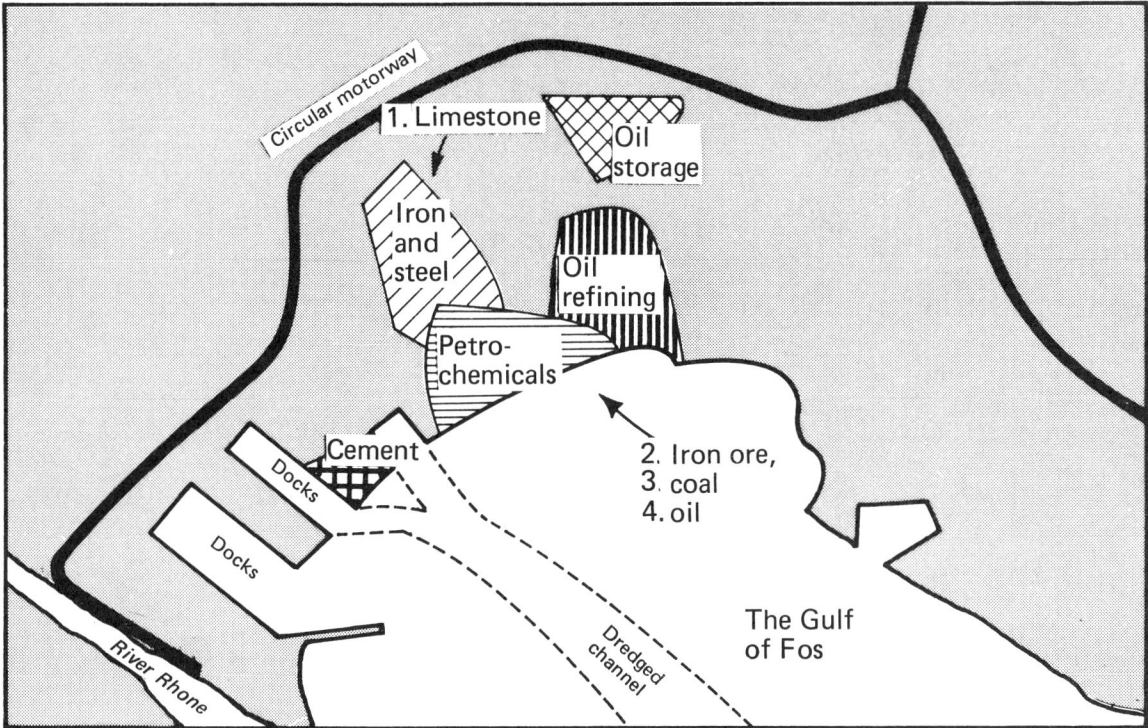

Fig. 97 The port of Fos

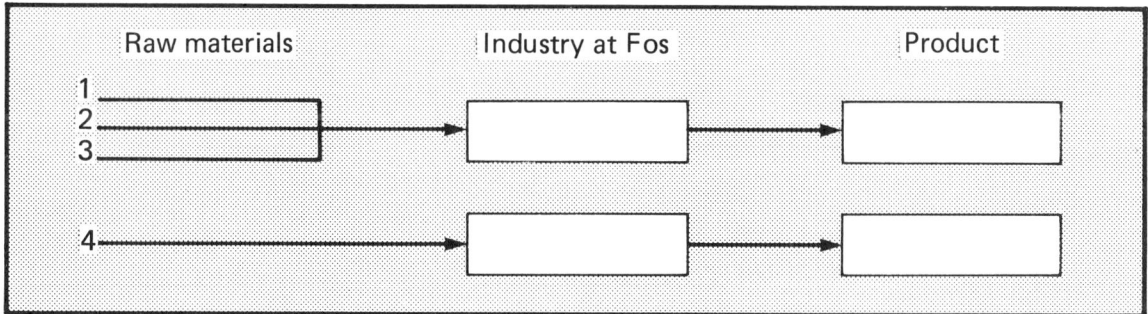

Fig. 98

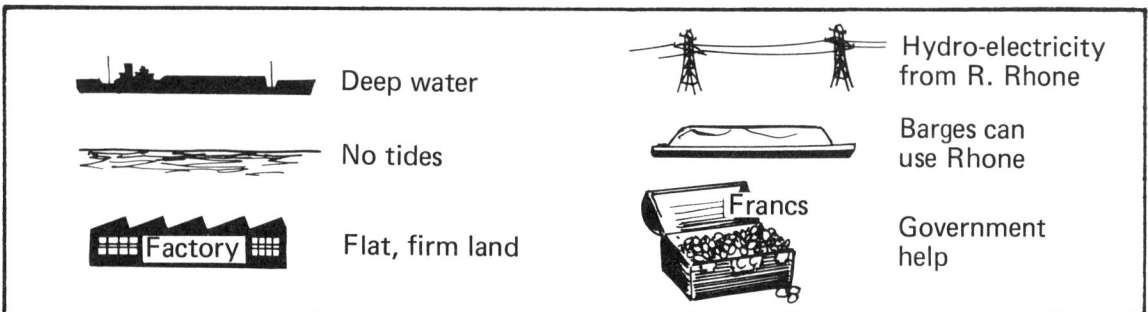

Fig. 99 Good points for Fos

4C The aluminium industry

Aluminium is a lightweight, silvery-white metal that is used in many industries. It has many useful features; great strength for its weight, easy to work, resists corrosion, conducts heat and electricity. Mixed with other metals it can form many different alloys. It is widely used in aircraft construction, the building industry and many consumer items, for example, refrigerators and saucepans.

1 Make a list of things in your home that have some aluminium in them.
2 Explain why aluminium is such a useful metal.
3 Fig. 100 shows the three stages of producing aluminium from bauxite. Use the diagram to help you write out and complete the following paragraph.

> The ore of aluminium is called _____ after the town of Les Baux in Southern France. After it has been mined it is partly reduced to a material called _____ before undergoing an electrolytic refining process which produces _____ . During manufacture there is a _____ loss of weight from every 4 to 6 tonnes of _____ to 1 tonne of _____.

Bauxite is mined in only a few countries. Some of the major producers are listed in Fig. 101.

4 Use your atlas to draw an outline map of the world and to mark on the countries which produce bauxite. Give your map a title.
5 Study Fig. 102. This table shows the major areas of the world that produce bauxite and aluminium outside of the USSR. Which three continents produce most bauxite? Do the same continents produce the most aluminium?
6 Which two continents produce the most aluminium? From which parts of the world will they obtain their bauxite?
7 Which country must buy all the bauxite it needs from other countries?

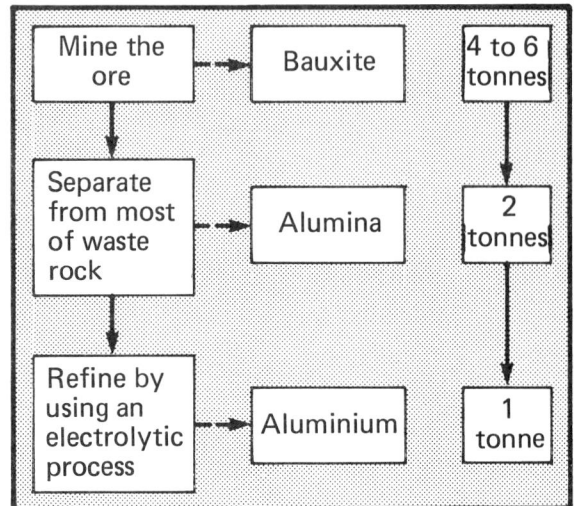

Fig. 100 Major processes in the aluminium industry

Country	1977 (million tonnes)
Australia	26.1
Guinea	11.3
Jamaica	11.4
Surinam	4.9
Guyana	3.3
India	1.5
Indonesia	1.3

Fig. 101 Bauxite extraction — major producers (excl. USSR)

	Bauxite (million tonnes)	Aluminium (million tonnes)
North America	2.0	5.4
South America	22.1	0.4
Africa	12.3	0.3
Western Europe	7.0	3.6
Japan	-	1.1
Australia	26.1	0.4

Fig. 102 Aluminium industry (excl. USSR)

In the last stage of the process of making aluminium large amounts of electricity are needed. This is necessary because the aluminium is finally separated from the impurities by passing an electric current through the molten metal. This method is known as electrolytic refining. Every tonne of aluminium produced requires 15 000 kWhs of electricity. That is enough electricity to power a one-bar electric fire for 90 weeks. Fig. 103 is a view along one of the smelters or 'potrooms' of the Norwegian aluminium smelter at Ardal. The smelter can make 185 000 tonnes of aluminium per year.

8   Explain what is meant by electrolytic refining.
9   How many kWhs of electricity would be used at the Ardal smelter in one year?
10 Look again at Fig. 103. Do you think an aluminium smelter has a large workforce?
11 Most aluminium smelters are located near a cheap source of electrical power. Why do you think this is so?
*12 Several of the countries listed in Fig. 101 are members of what is called the *Developing World*. This means that they are only slowly developing their economies and raising standards of living. How do you think these countries would benefit from mining bauxite?

Fig. 103

4D Aluminium in Norway

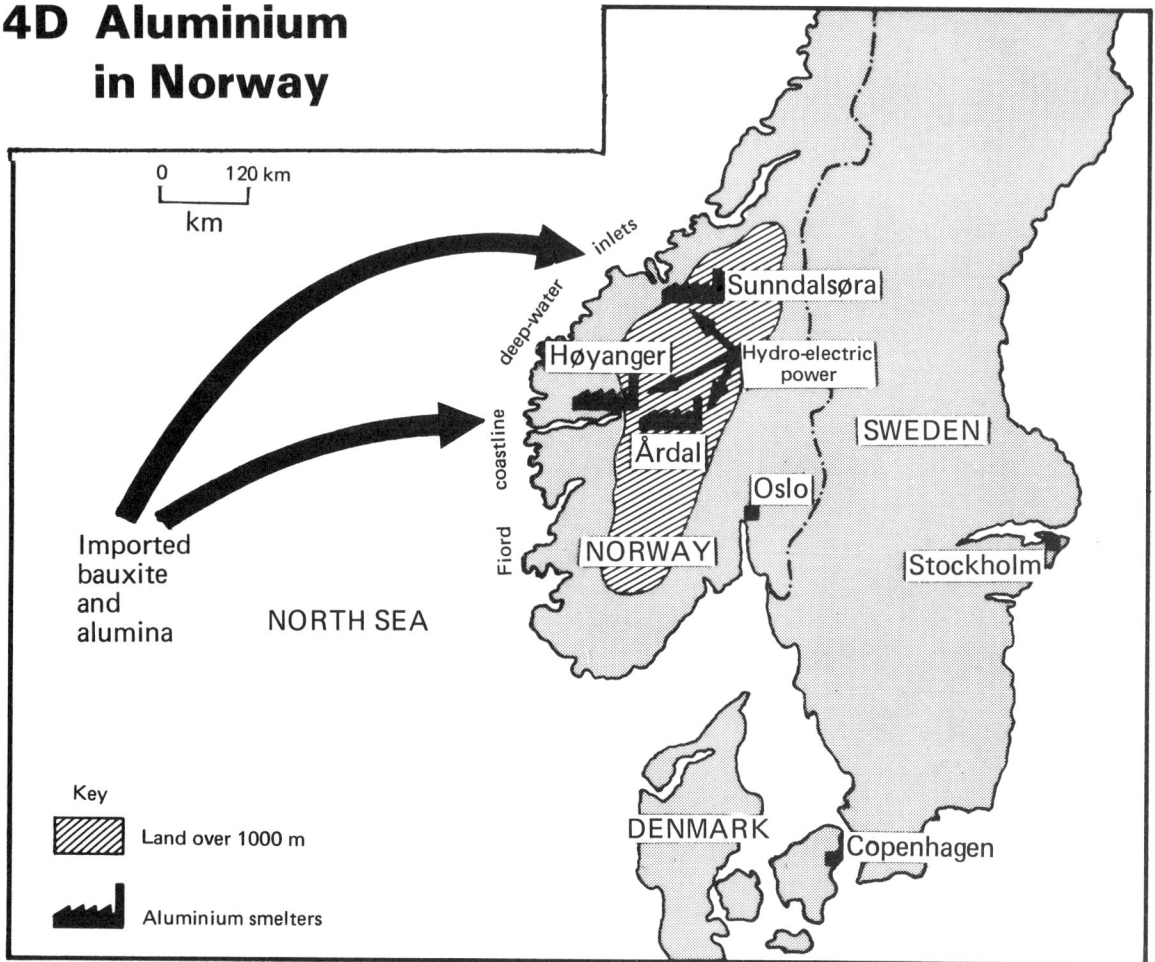

Fig. 104 Aluminium in Norway

The manufacture of aluminium needs large amounts of electricity. Therefore one of the main factors in locating an aluminium smelter is cheap electricity nearby. In Norway the ASV Company operates three smelters in the south of the country. The position of these three smelters can be seen on the map in Fig. 104.

1 Name the three aluminium sites.
2 Each works is located on a deep-water inlet of the sea. What are these inlets called?
3 What raw materials are imported by the ASV Company for the smelters?

Fig. 105

Fig. 106

One source of electricity in southern Norway is hydro-electric power (HEP). This form of electricity is generated by the power of falling water driving turbines linked to generators.

4 Look at the map in Fig. 104. Why do you think the high land of southern Norway would help in the making of hydro-electric power?
5 Fig. 105 shows the typical location of an HEP station in Norway. Draw a labelled sketch of the photograph to show: *highland, pipes carrying water to HEP station, HEP station, lake.*

Figs. 106 and 107 show the position of Sunndal Verk, one of Europe's largest alu-

minium smelters, at Sunndalsøra. It produces over 300 000 tonnes of aluminium a year, mostly for export to West Germany, Sweden, Britain and the USA.

6 Study Figs. 106 and 107 that show the Sunndalsøra aluminium smelter.
   (a) Describe the location of the works.
   (b) Describe the site and layout of the buildings.
*7 Draw a sketch map of the Sunndalsøra works to show: *position of smelter, flat land, quay, fiord, source of power.*
*8 HEP stations and aluminium smelters are often built in areas of beautiful scenery. Do you think local people would like them built in their area? Give your reasons.

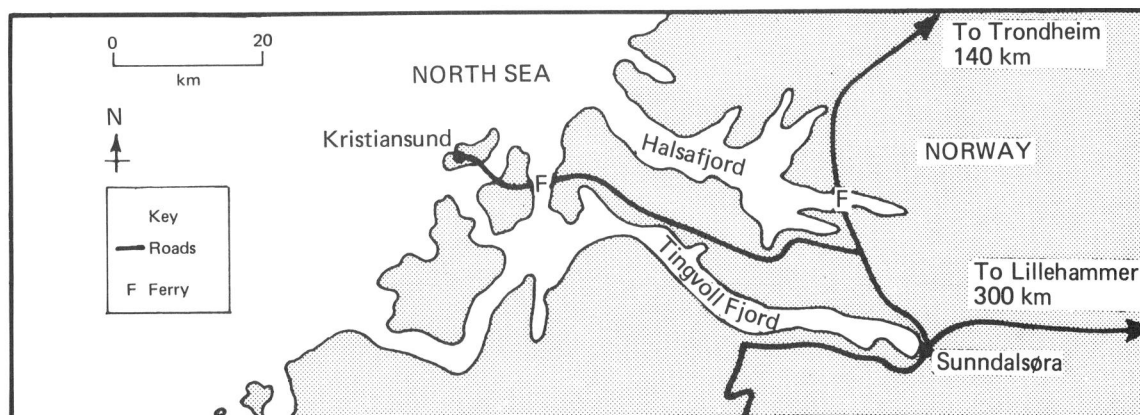

Fig. 107

55

4E Industry on the move

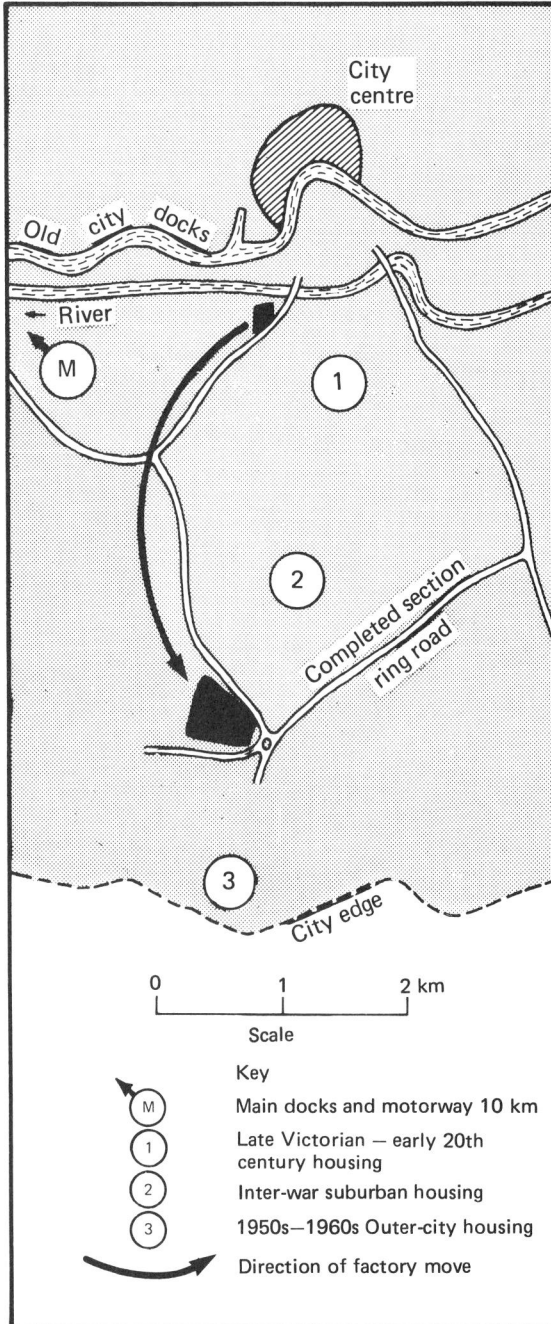

Fig. 108

Between 1969 and 1975 a large firm in a British city moved its factory from a site near the city centre to a new site in the suburbs. The move cost about £20 million and 4000 staff were transferred to the new site. Let us find out why the firm moved.

The factory near the city centre had been built in 1886 and had seen over 80 years of continuous use. The factory contained a complex production process involving much machinery. But new machines were becoming available and the buildings were unsuitable for the latest technology. The firm used road transport to bring the raw materials from the docks and to distribute the finished product. In the 80 years of the factory's life the traffic around the site had increased many times. These factors were making running costs higher than they would be in a new *purpose-built* factory, that is a factory specially designed for the new technology. The firm considered three choices of site when the move was being planned.

	Advantages	Disadvantages
Choice One – move to another city	Receive grants of money from the government.	Lose a trained, experienced and loyal workforce.
Choice Two – move to a site near the existing factory.	Little extra travelling for workforce.	Low-lying site with risk of flooding. Poor road access.
Choice Three – move to city suburbs	Large land area available from City Corporation. Better road access.	Longer journey to work for employees.

Decision → Choice Three

Fig. 109

Fig. 110

1 What were the two main reasons for the factory move?
2 Look at the map in Fig. 108. Describe the area in which the 100-year-old factory was situated.
3 Study Fig. 109 which shows the choices faced by the firm. Explain what you think was the main reason for the firm deciding on Choice Three.
4 How far did the firm, in fact, move?
5 What is meant by a 'purpose-built' factory?

6 Look at Fig. 110 and Fig. 111 which are views of the firm's old and new factories. Describe the appearance and position of the two buildings.
*7 Investigate the movement of firms in your town or city. Have any firms moved to new sites? Why? On a map of your town, plot the moves and see if there is any pattern.
*8 Increasing use of road transport has had a great effect on the location of factories in recent years. Is this true in your town? If there is a motorway near you, investigate its influence on recent factory buildings.

Fig. 111

4F Industry in towns — trading estates

Dairies — bottle stores	Timber merchants
Engineering	Body repairs
Builders merchant	Fire station
Garage/service station	Glass and glazing contractors
Heavy transport/waste disposal	Corporation cleansing
Hardware wholesaler	Container transport
Commercial vehicle hire	Building contractors
Manufacturer of industrial controls	Wholesale grocers
Electrical components — distributors	Carton packaging
Clothing wholesaler	Hardware and tool distributors
Tyre distributor	Plastic stockists
Business equipment — wholesaler	Photographic wholesalers
Industrial tools — wholesaler	Boat manufacturers
Caravans — retailer	Hose equipment suppliers
Motor car show rooms	Pet shop suppliers
Service station	

Fig. 112 Trading estates — firms and services

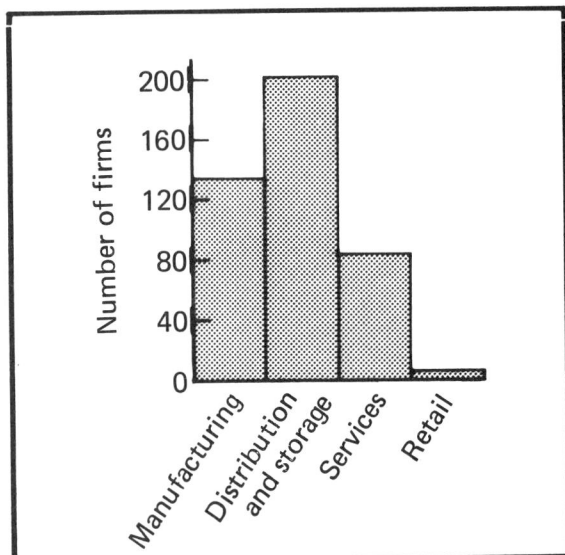

Fig. 113

Trading estates, sometimes called *industrial estates*, are areas in towns made up of factories and warehouses. They are found in most large towns and cities in Britain. In most cases they have been carefully planned and are laid out with factories and warehouses specially designed to serve many uses.

Look at Fig. 112. This is a list of firms and services to be found on one of the twenty trading estates in and around the City of Bristol.

1 Using the headings below count the number of firms in each group.
   — *Distribution and Storage (wholesalers)* (firms that store products and distribute them to retailers and the general public)
   — *Services* (help or skill sold to other firms and the general public)
   — *Retailer* (direct sale to the public)
   — *Manufacturers* (firms that make a product)
2 Which group has the most firms? Can you suggest reasons for this? It may help you to know that Bristol is well positioned in relation to the road network, particularly motorways which cross the West Midlands and the south-west of England.

We have only considered one trading estate; a more accurate picture is seen if more areas are included. Fig. 113 shows the types of firms found in nine chosen trading estates in Bristol.

3 How does this bar graph compare with your answers to Questions 1 and 2?

Look at the map, Fig. 114. This shows the main features of the site of the trading estates which lie two miles south of the centre of Bristol.

4 Using the information on the map and the photographs (Figs. 115 and 116), describe the site of the trading estates. Consider roads, residential areas and access.

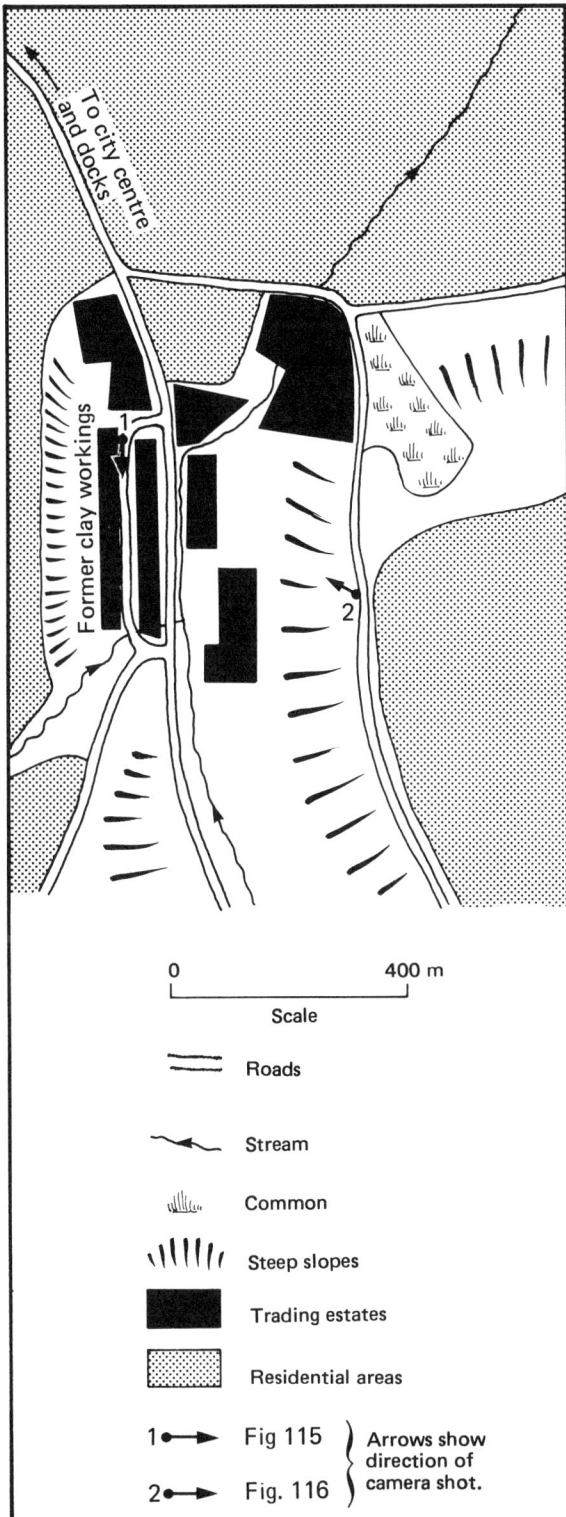

Fig. 115 Road through trading estate

*5 *Project idea*

Conduct a survey in your own town of industrial areas and trading estates. Use a simple grid as shown below to record information.

Name of trading estate . . .			
	Name of firm	Activity e.g. wholesaler	Type of building
1			
2			
3			
4			

*6 *Other activities*

Draw a simple sketch map showing the layout of the industrial area/estate. Collect newspaper cuttings about industry in your town. Contact your local planning department and ask them for help. They may have made a survey as well.

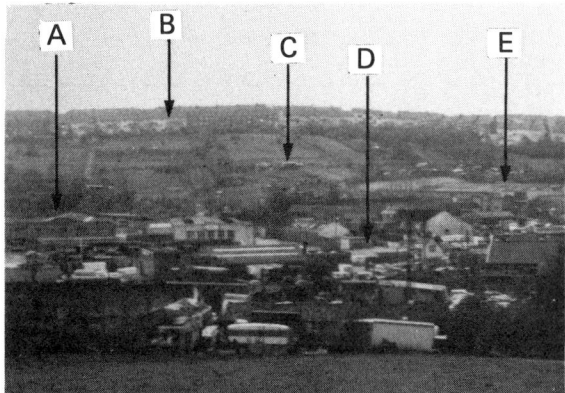

A old clay workings   D main road
B residential area   E engineering works
C allotments

Fig. 116 View of trading estate

Fig. 114 Simple sketch map of firms listed in Fig. 112

Map legend:

0 — 400 m
Scale

Roads
Stream
Common
Steep slopes
Trading estates
Residential areas

1 ●→ Fig 115 ⎫ Arrows show
2 ●→ Fig. 116 ⎬ direction of
          ⎭ camera shot.

To city centre and docks

Former clay workings

1

2

59

5

Tourism

Fig. 117

5A The growth of tourism

Figs. 117 and 118 show very different holiday scenes. Fig. 117 was taken in Benidorm, just north of Alicante, on the Mediterranean coast of Spain, while Fig. 118 was taken in the Austrian Alps. They represent contrasting holiday styles.

1   Find Alicante and the Austrian Alps in your atlas and show them on a sketch map.
2   Write a paragraph to say what is attractive about each holiday photograph.
3   Follow this with another paragraph to say what you might not like about the type of holiday shown in each photograph.

Fig. 118

The holiday industry has grown rapidly over the past 20 years for a number of reasons. People have more money to spend. They have more holiday time. More people own cars. Holiday costs have been kept down by *package holidays.*

A package holiday is offered by a travel company which makes all the arrangements for a holiday at a popular resort, including travel and accommodation. By this means the company can *charter,* or hire, a whole aircraft at a cheap rate and be sure of filling it with passengers. In the same way cheaper rates can be offered by hotels as they are sure they will be kept full.

The photograph in Fig. 119 shows that not all holidaymakers look for summer sun. More and more people choose a winter holiday.

4   What sporting activity is shown in the photograph?
5   The photograph shows several of the things needed for this sport to be successful. List as many as you can.
6   With the help of an atlas name four countries in Europe important for winter sports.
*7   Make a list of the kinds of job you would find in the three tourist centres. Which jobs would be found in all three places?
8   *Tourism* is now a major industry. Is it primary, secondary or tertiary? Explain your answer.

Fig. 119

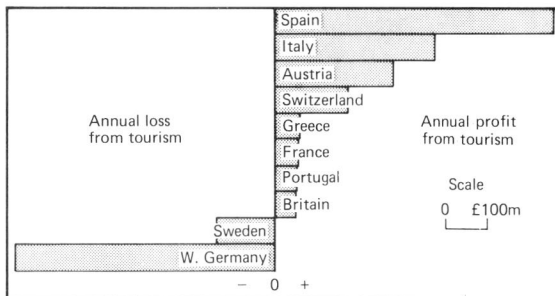

Fig. 120

Fig. 120 shows the profits and losses concerned with tourism for selected European countries. The columns are measured in millions of pounds.

9   Work out the value of the columns for Spain, France and Germany. Make sure you say if each is a profit or loss.
10   For the countries shown on the bar graph name those with a Mediterranean coast. Use an atlas to help you.
11   Which range of mountains is shared by Austria, France, Italy and Switzerland?
12   What do you notice about the position on the bar graph of the groups of countries mentioned in questions 10 and 11?

Climate figures for Palermo, Sicily:
Average temperature   coldest month (Jan.): 10°C; hottest month (Aug.): 25°C
Rain   annual: 97 cm; wettest month (Dec.): 18 cm; driest month (July): 0.6 cm
Each year   rainy days: 82; frosty days: 1

13   Use an atlas climate map and the climate figures above to write a paragraph describing the climate of the Mediterranean coast.
14   Look back at Fig. 117 and add to your paragraph another attractive part of the weather not shown by the climatic figures in question 13.

In 1981, 100 million people spent their holidays around the Mediterranean Sea. By the year 2000 it is estimated that this figure will have reached 200 million holidaymakers a year.

*15   Write an essay describing in your own words some of the advantages and disadvantages for Spain of so many people crowded together along its shores for a few weeks of the summer.

5B European holidays

The map in Fig. 121 shows how many people from Britain and West Germany visit some European countries each year.

1 Work out the total numbers of tourists to these countries from Britain and West Germany.
2 How many million tourists does Spain receive from Britain and West Germany?
   What percentage of Spain's total is that? Is it a higher or lower percentage than that for Italy?
3 Name the countries visited by large numbers of West German tourists each year.
4 Nowadays people are prepared, and can afford, to travel further for their holidays. How far is it in a straight line from London to (a) Benidorm and (b) the Austrian Alps?

This extract taken from *The Sunday Times* shows some of the effects of tourism on the Mediterranean.

The Sunniest Sewer in Europe

Besides being the world's largest swimming pool, the Mediterranean is rapidly becoming the world's least flushed sewer. Every year millions of tourists flock in to swell many coastal communities to 10 times their usual size. But how much longer will they tolerate oil-stained towels and dysentery.

To solve the sewage problem would cost around £2500 million and take at least 15 years. There is a conflict of interests, for while Spain and France are anxious to nurse their beaches back to their former glory, Algeria and Libya are impatient to exploit their oil and natural gas fields by building new tanker terminals.

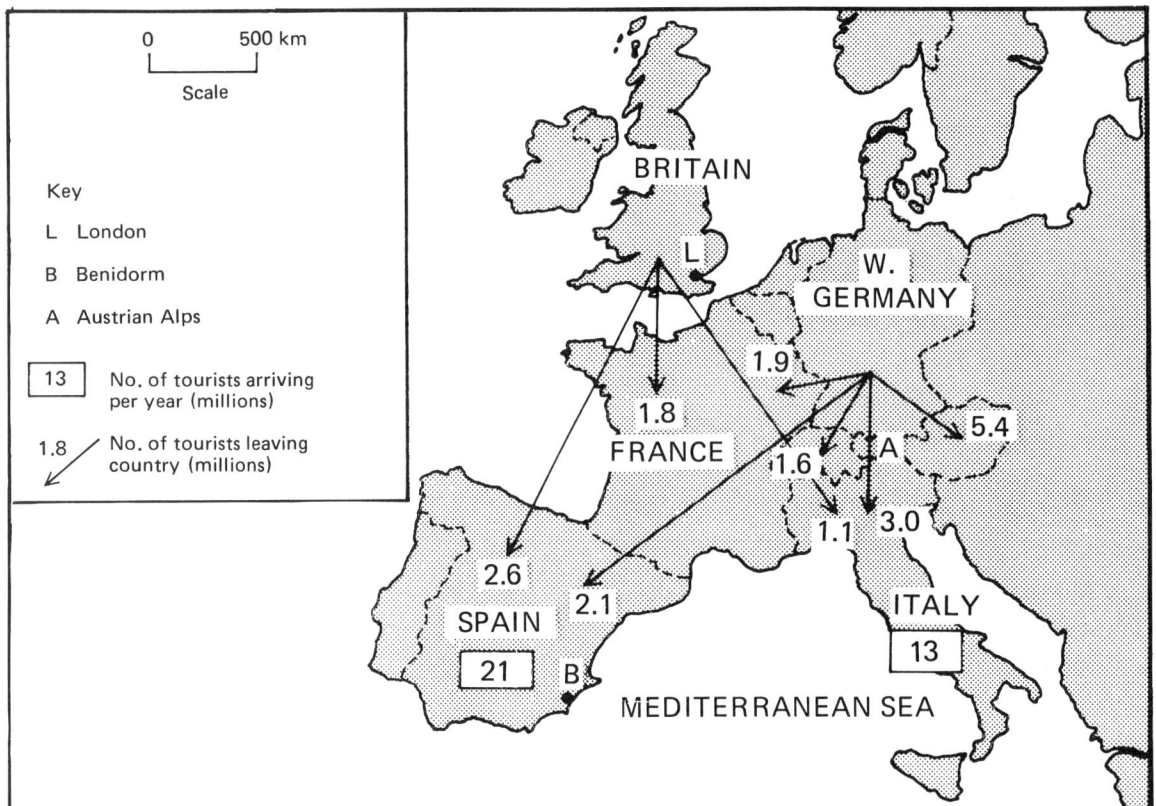

Fig. 121

5 Look at Fig. 122 which shows a *polluted* beach in Italy.
   Describe the scene in your own words.

The map Fig. 123 shows which Mediterranean coasts have the most serious ***pollution***.

6 Name three causes of pollution.
7 Which two countries have the most polluted coasts?
8 Why are the northern shores of the Mediterranean more polluted than those to the south?
*9 Look at an atlas map of the Mediterranean to help you explain why the Mediterranean has hardly any tides and takes about 100 years to completely flush itself with new water.
*10 Write an essay describing some of the ways in which Mediterranean pollution could be reduced in the future. Also outline some of the difficulties and problems that must be overcome before this can happen.

Fig. 122

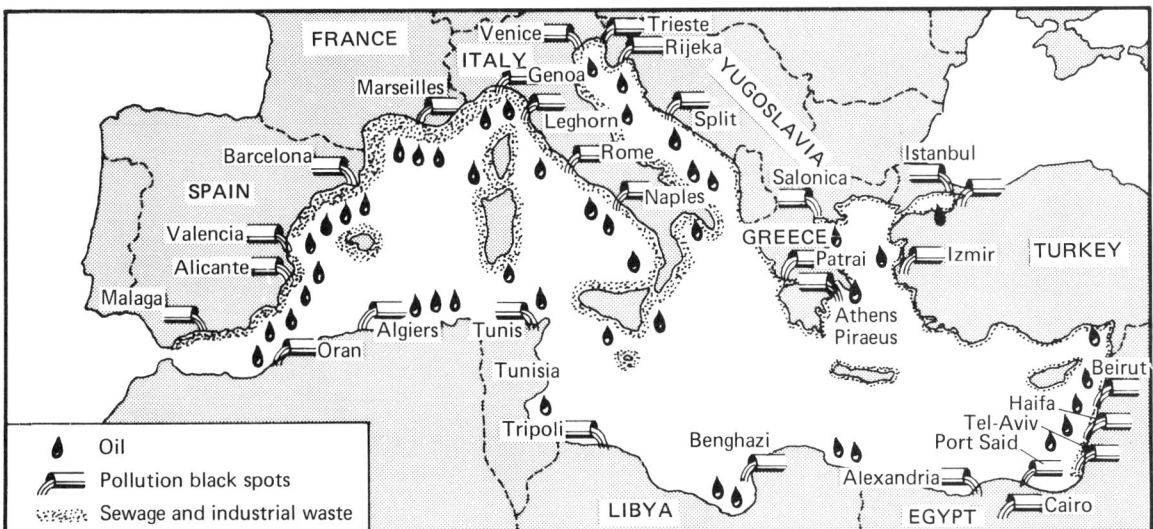

Fig. 123

Glossary

assembly line — a product is put together piece by piece as it goes along a moving belt

break of bulk — the place at which goods are changed to another form of transport for wider distribution

bulky — heavy, loose cargo, low in value but needed in quantity

coke — fuel made by heating coal with little air; used in the making of iron

cottage industry — small-scale manufacture carried out in people's homes

density of population — the number of people living in a given area; if a fixed area is used (e.g. km^2) one place can be compared with another

end-product — the final manufactured product

exports — goods sold to a foreign country

hardwood — a close-grained timber; our deciduous trees are hardwoods

heavy industry — industry that uses much heavy material and needs a lot of handling by machines

imports — goods brought into the home country from a foreign country

Industrial Revolution — the growth of the factory system in the 1800s based on the steam-engine and coal power

industry — organised work, manufacture or trade

inputs — materials, power and anything else put into an industrial process

integrated works — all the manufacturing stages of a product take place as a continuous process at a single works

light industry — industry that uses small quantities of lightweight, easily-handled materials

manufacture — to make something

mass production — the manufacture of large numbers of the same end-product

open-cast mining — the mineral is dug out after covering rocks and soil have been removed

ore — rock containing a valuable mineral

outputs — the end-products of an industrial process

overburden — the rock cover above the mineral in an open-cast mine

papier maché — a mixture of paper and paste that sets into a hard material

pollution — the spoiling or harming of an environment by the dumping of our waste products

primary industry — industry which makes use of the natural products of the earth and provides other industries with raw materials

pulp — the broken-down fibres of wood

quick-freezing — factory food preservation using a blast of very cold air

raw materials — materials which are made by industries into things to sell

secondary industry — industry which makes products from raw materials

service — nothing is made; the customer pays to have a task completed

slurry — a mixture of fine mineral particles in water

social services — public services not paid for directly by customers but through everyone's taxes, etc.

softwood — the timber of coniferous trees; used a lot in the building industry

tertiary industry — industry that does not make an end-product; payment is for a service given

tourism — the industry that caters for holidaymakers

trading estate — an area, with good roads, having many light industries and service trades